C0-ALA-045

Peeling the Sweet Onion

Unlayering the Veils of Identity and Existence

By
Martin E. Segal

Author of THE GURU IS YOU

Portions of Part VIII are reprinted from *AIDS: From Fear to Hope*, Copyright 1987 by New Age Publishing Co., ISBN 0-934619-02-6.

This is another book for Personal Growth and Transformation From

NEW AGE PUBLISHING CO.
P.O. Box 01-1549
Miami, Florida 33101
(305) 534-8437

Original Cover Artwork by: David Ellis Rubinson
Typing by: Debra Daniel
Layout and Design by: Dark Horse Press
Printed by: BookMasters, Inc.

Second Edition—1991
Library of Congress Catalog Card Number: 90-091503
ISBN: 0-934619-03-4

RIPPLES

Ah! So
Be still and know
Accept and let go
Ebb and flow

All forms arise,
All pass away,
No place to go
Nowhere to stay

We resist and avoid,
We grasp and we cling,
Our precious dramas
The ten thousand things

The past that we hold,
The future that we fear
Life drips away
Year after year

Searching for happiness,
We want to know how,
The answers lie inward
The perpetual now

Layers of self,
Strip them away,
Emptying the fullness
Seizing the day

Coming back home,
Where it began,
Filling with Emptiness
The true cosmic plan

Ebb and flow
Accept and let go
Be still and know
Ah! So

RULES FOR BEING HUMAN

1. You will receive a body. You may like it or hate it, but it will be yours for the duration of this particular lifetime.
2. You are enrolled in a full-time informal school called life. You will learn lessons. Each day in this school you will have the opportunity to learn them. You may like the lessons or think them irrelevant and stupid.
3. There are no mistakes, only lessons. Growth is a process of trial and error experimentation. The "failed" experiments are as much a part of the process as the experiment that ultimately "works."
4. A lesson is repeated until it is learned. A lesson will be presented to you in various forms until you have learned it. When you have learned it, you can then go on to the next one.
5. Learning lessons does not end. There is no part of life that does not contain lessons. If you are alive, there are lessons to be learned.
6. "There" is no better than "here." When your "there" has become a "here," you will simply obtain another "there" that will, again, look better than "here."
7. Others are merely mirrors of you. You cannot love or hate something about another person unless it reflects to you something you love or hate about yourself.
8. What you make of your life is up to you. You have all the tools and resources you need. What you do with them is up to you. The choice is yours.
9. Your answers lie inside you. The answers to life's questions lie inside you. All you need to do is look, listen, and trust.
10. You will forget all this—over and over again.

ACKNOWLEDGEMENTS

"Life is like an onion—You peel if off one layer at a time, and sometimes you weep."

Carl Sandburg

After one has traveled for a time on the Path of Awakening, accumulating knowledge and experience, insight and awareness—one has a sense of naturally "daring to be different" by automatically taking a step forward away from consensus consciousness to go beyond the boundaries it arbitrarily imposes. It also becomes obvious that everything in life is the curriculum for learning our important lessons of personal growth. The distinctions between specific bodies of information merge into one fluid mass of living truth. So rather than thank any one teacher or body of knowledge or learning experience, it is more appropriate that I acknowledge with deep appreciation the Journey itself in its awesome moment to moment perfection.

Special recognition must however be given to Ken Keyes and Ram Dass and their teachings. They provide continuing inspiration in linking the practical and theoretical aspects of personal growth and higher consciousness.

DEDICATION

To that greater part of myself that exists beyond my physical world of forms; sitting back behind it all as an interested observer watching with such loving compassion as I get lost in my dramas, find my way out again, submerge, come back up for air, rise and fall, remember and forget, come and go, over and over and over again. It is my loving partner in the dance of life.

CONTENTS

INTRODUCTION

"Have you been doing any writing lately?" That's what Ken Keyes (million selling author of "Handbook to Higher Consciousness" and other personal growth books) said to me when I bumped into him that Sunday evening down in Coconut Grove. He was on his way to a World Peace Conference in Costa Rica, and I was out for an uncharacteristic Sunday night stroll trying to sort out the many thoughts and ideas for a new book that had been running through my head the past few months. I hadn't seen him in years and he was 3,000 miles away from home. Nevertheless, synchronicity had struck and here we were. At that moment I realized that it was time to get started with the birthing process by putting words down on paper.

A period of gestation had been ongoing. I had been receiving subtle urgings from my inner voices and not so subtle ones like this meeting with Ken. Since writing "The Guru is You" in 1985, my personal journey of awakening was of course continuing. I had been, now that I reflect on it, inputing much additional information, gaining new insights, experiencing, incubating, and preparing to build upon prior foundations.

"Guru" was a personal catharsis, a sharing of teachings and a learning experience all rolled into one. The journey had started for me 10 years ago when I began to consider the possibility that the perceived external negative world of fear and isolation, to which I was a card carrying member, entitled to all the attendant disadvantages of anxiety, fear and unhappiness — was not the "real" world at all. Rather, the limitations of my own mind's patterns of thought and my attitudinal reactions to situations created my view of the Universe.

I was looking everywhere for fulfillment except the most obvious place—within myself. I was ever striving to grab the brass ring of happiness that seemed "out there" just beyond my reach. I had lost my way, sucked into the societal conspiracy that turns us inside-out and deludes us into believing we are "less than." But gradually I became aware that I always held that brass ring within my hand—I had never let it go. My opening to the greater possibilities of existence reached a point of organized presentation five years later with publication of "The Guru is You."

Responses from many of its readers in the form of letters, phone calls, and the blue verdict card mailers that were inserted in the book were most gratifying. There seemed to be a common recognition of the poignancy of our shared predicament of struggle to find an experience of lasting happiness and peace of mind in this complicated world in which we live. My suggestions that our problems were essentially self-created due to the manner and methods in which we reacted to the learning experiences of our daily lives, our erroneous external focus of awareness, and the nature of the layering process by virtue of which we learned to become dysfunctional by obliterating our inherent inner wisdom—struck a responsive cord. They were intrigued, as was I, by the metaphor of a "mummy" that is covered by increasingly dense layers of negative "stuff" which is learned from the external focus of an unconscious life. The more layering, the less vision, mobility, and flow of the joyful energy of a happy life, until a basic misperception is realized: perhaps we were already enlightened and possessed all the insight we needed for living a happy life, but just had lost our way and had forgotten where and how to find it. Our problem was not lack, just location!

I was urged to further develop and refine these concepts that help provide us with answers to our perennial questions: (1) Who/ what am I and where did I come from? (2) Why am I here? (3) What are the rules of the game? (4) What is the nature of suffering, and how can it be alleviated? (5) What is a life well lived? (6) What happens when the game of life is over?

An exploration of possible answers to these questions would furnish insights to better focus our eyes of understanding so that the pictures of our game of life would become more vivid and it's playing more pleasurable. They would help us to pierce the veils of identity and existence that were clouding our vision. Our attitudinal healing required a new map to guide us home.

In order to be a cartographer, one must be well traveled. In my five years on the road since "Guru" much additional knowledge had been gained. I studied eastern and western philosophy, their psychological counterparts, and related reference sources; attended lectures; participated in workshops; facilitated New Age discussion groups; listened to audio tapes; watched video tapes; read numerous books of non-fiction and fiction in these areas of personal growth and higher consciousness; had channeling experiences; and most importantly became more aware of my everyday learning experiences by living an active life in the market place. Some of these experiences were gentle, and others remarkably intense. This was the arena of opportunities for growth and awakening.

The inner classroom of dreams, meditation, contemplative experiences and reflective thought had also been valuable and important. Being both an active participant in the game of life and a detached observer of it's intricacies can be most illuminating. But this is not a sequel to "Guru"; rather, it's a continuation of the previous learning and sharing process containing new insights that move me farther along the path of enlightenment.

The result is a synthesis and codification of the most important aspects of these divergent sources of wisdom presented in a conversational format which is comprehensive, easy to understand, and practical in application. It has been stripped of esoterica and mysticism, so that it is pure and easily digestible, appealing to both the intellectual and the pragmatist.

The ideas presented in this book are basic and straightforward, reflecting both their simplicity and their profound importance to a life experience of positive growth, peace of mind, and lasting personal happiness. They trace the story line of our common soap opera.

We birth as a perfected miniature being that is pure loving kindness, openness, trust and joyful moment to moment awareness connected to its divine origin. The subsequent filling of this perfectly empty vessel with societally acquired layers of negative belief systems, fear, limitation, and judgement, creates a sense of personal isolation and separateness from its greater identity; the spacious awareness from which it came. This societal layering process creates negative models of belief and behavior by teaching us to exclusively focus upon the exterior world as experienced through our physical senses, to identify with it and become absorbed in its heavy dramas as "real." These constricting layers cause a disconnection from the underlying positive creative essence that is our true inner self, and we suffocate in our illusions with a life of much pain and suffering.

The play can nevertheless become a more mellow-drama through a process of unlayering where we discreate unproductive patterns of thought and behavior by turning our world of focused awareness outside-in, and returning to home base. We cleanse ourselves by utilizing various methods to empty out these acquired toxins, and we begin the important reconnection with those greater aspects of our awareness that had previously been neglected, closed down and lost. We are then able to refill ourselves again in a more conscious manner and recreate a more enjoyable and productive life that incorporates more of the desired aspects of enthusiasm, effectiveness and positive fulfillment.

These changes in focusing the clarity of our lenses of perception, reversal of our external awareness stance and the retracing of our steps back "home" can enable us to maximize the positive experience of our life's journey. We become aware of how we had shut ourselves down into minimally functioning robot-mummys, covered from head to toe with dense layers of acquired negative patterns of thought. We have been staggering around in the dark, looking at our world through dense bandages of doubt, limitation, and fear that reflect our shared perception of exterior consensus reality.

We realize that our approach has been wrong. We were looking for answers in the wrong places. We were lost, trying to find our way back "home," when in fact we really never left it in the first place. Our flashlight was shining in the wrong direction; no wonder we couldn't see clearly enough to find our way.

These important concepts of layering of negative thought patterns; unlayering, and discreation of unproductive learning; relayering and recreation of desired realities; and the *outside-in* methodology are all presented in detail. Also presented in depth is an examination of our true identity; the purpose of our existence and its aspects of happiness and suffering; the fundamental natural laws of the universe that really govern our journey; practical strategies for success; a look at the concept of death; and a prognosis for the coming decade as we enter into the 21st century.

In creating the format for this book, I have blended psychology's study of human behavior with spirituality's focus on the greater aspects of existence—a psycho-spiritual viewpoint. The intellectual structural foundation is first presented as a sort of black and white outline of knowledge. It serves as a learning bridge of awareness which can be crossed experientially through affirming events we learn to practically generate in living color as we turn our world *outside-in*. These everyday experiences bring the theoretical framework to life and give it practical meaning. Having first absorbed the information by intellect and secondly affirmed it by experience, we can then proceed to take this elevated consciousness and expanded awareness "out there" into the market place to change our lives and those of our community of fellow beings for the better.

PROLOGUE

This is not a cookbook, at least not in the traditional sense, but it certainly contains food for thought. It combines many different ingredients that I've tasted in my travels of dining at various restaurants of personal growth. There are heaping tablespoons of Eastern Philosophy, a teaspoon or two of Theology, large servings of popular Western Psychology—the self-actualizing humanities of the 60's; the cognitive, rational emotive concepts of the 70's; Gestalt therapies and group processing; Transpersonal approaches of the 80's; a dash of quantum physics; and liberal helpings of Metaphysics.

It has been folded together in a delicate blending, given time to marinate, and then baked in the oven of human experience. The finished product is a psycho-spiritual combination of all its components, and much more.

Why the title, "Peeling the Sweet Onion"? We associate onions with tears, and that is our experience of external reality for much of our lives until the unlayering process begins and we discreate the negative patterns of thought that cause our suffering. But as we turn our focus of awareness *outside-in*, we can shed the veils that impede us, peel them away layer by layer, and expose the sweet center of our being where our inherent joy and wisdom resides.

Parts of this book may be familiar as they reinforce existing knowledge. Other parts may remind you of forgotten truths. In either event you will probably agree or at least be comfortable with the possibilities that are suggested. However, there may be other material that is at odds with your current mental models of truth. You may notice your initial reaction of disagreement or rejection. If and when this occurs please do us both a favor. Open yourself to the possibility of new truths. Play the game of "considering the possibility" that this may be the Way things work after all. If you open yourself to this extent and allow the information to at least come in I guarantee you won't be disappointed. This is a guarantee of increased insight and enhanced awareness far beyond one of monetary considerations.

There are some inherent difficulties in presenting this material. The most basic is the necessity that we communicate with words. By their very nature, as symbolic representations of concepts communicated through our thinking minds, words are a part of our world of form for which we have established boundaries of interpretation according to our physical senses. Since we will be talking about aspects of a multilayered spectrum of consciousness that exists beyond this physical world of forms, descriptions of the invisible world using the word terminology of the visible world may at times be difficult to conceptualize. However, for the sake of clarity, the attempt has been made to keep things simple.

A similar problem is the fact that this material is presented, again for the purpose of clarity and the ease of comprehension, in a linear and sequential fashion, like a logical outline. Time and space are concepts of the world of form. The invisible world of our conscious awareness, out of which the forms are created, exists beyond form but must also be described like the world of form in this step by step manner; notwithstanding its inherent timeless and infinite nature.

The material is presented in a simple and direct manner. There is no attempt to dazzle with secret rituals, spiritual mumbo-jumbo, or occult symbology. It's not necessary. Within the simplicity of the concepts lies their deep and profound meanings.

Everything presented has practical application in your everyday life. The theoretical concepts build the bridge for one to cross into affirming experience. The same balancing that is necessary to understand and effectively use these concepts is provided in their presentation because the ultimate objective is, after all, to incorporate them into one's personal game plan to more fully experience a life of positive growth, joyful living, and inner peace.

This sharing of my understanding of the nature and meaning of The Game of Life, in all its variegated aspects, comes to you as it must from my current vantage point. Like the narrator in James Hilton's "Lost Horizons" who spent a large part of his life searching for Shangri-La, once it is located the quest really begins in earnest. Within that fictional city, hidden behind the dense clouds and sheer slopes of the Himalayas lies his true spiritual home: an ever changing landscape that is a kaleidoscope of subtle nuances within its veils of multi-dimension.

After forty years of wandering in the desert of consensus unconsciousness, lost without knowing it, I became aware of my predicament and, with the best map I could find at that time, started the long Journey back home, to find my own Shangri-La, lying within the great valley of awareness and insight all around me. I look out at its broad expanse, and what follows is what I see!

OVERVIEW

THE STAGES IN THE JOURNEY OF AWAKENING

Our true "home" is the realm of the macrocosm, *moment to moment choiceless spacious awareness,* the formlessness from which all form comes, our underlying oneness, our greater identify, the greater part of our being. It is beyond time and space, perpetual, infinite and everlasting; neither having a beginning or an end—it just is, the ultimate beingness.

It is the Universe with a capital U, and its compositional elements are unconditional love, kindness, compassion, total acceptance, non-judgmental allowance, joy, cosmic humor, ecstacy, rapture, and all the other adjectives which can not describe in words this undescribable "all that is" that is beyond words.

It is pure energy awareness, where there are no boundaries and a state of total interconnectedness exists. This is the realm of divine perfection in the sense that the Universe is the shared consciousness that runs the whole game of life! There is a plan which allows a place for everything and at the same time keeps everything in its proper place. There is no limitation or lack. It is total benevolence and wisdom and peace in an ever flowing and merging collage of energy.

Within this ocean of universal energy there are the individual waves that are its microcosm. These are the soul sparks of pure conscious awareness that are aspects of all that is—separate yet combined like the grains of sand in a vast desert and the drops of water in the cosmic ocean—interacting with each other on the multiple levels of universal consciousness in a world beyond the physical senses.

In PHASE ONE the soul essence decides to lower its vibrational level and take a physical birth in the dense world of forms that is the earth plane experience. This is not a random choice, rather the soul essence has been reviewing and digesting the results of prior visits (incarnations) to the three dimensional physical plane to ascertain the degree and extent of learning its various lessons for personal growth and continuing creativity (evolution) which is the inherent purpose of all consciousness. The earth plane is the richest and most graphic classroom for learning because of the incredible variety of forms, their sticky attractiveness, and the seductive atmosphere in which they interact. The spiritual entity chooses to take birth in order to be provided with the optimum curriculum and atmosphere for learning so that it can use the stuff of the world—the world of forms, as its vehicle for awakening. This awakening process is much faster on the earth plane because of the intensity of the game of life played there.

When soul energy awareness decides to take its trip passing through a particular physical lifetime, it creates at the phase one level, a roadmap of major learning experiences to be encountered, decides whether the trip will be taken as a male or female, picks out its physical and personality characteristics, (the space suit which will be worn during the trip), chooses the appropriate astrological, parental and geographical environment to facilitate the blueprint, and then what we call physical birth takes place.

In the beginning we retain the connection to our underlying true identity which is the greater part of our being. It works like an umbilical cord of knowing awareness that tracks the spiritual silver cord, maintaining the necessary spark of life and universal wisdom in this physical body which we inhabit as our traveling vehicle for this trip. For awhile we maintain awareness of the overall game and its multiple levels. However, the earth plane is a tremendously rich training ground and very seductive. In order to travel we are inhabiting a form (our space suit) which must now navigate in this world, and we begin to lose our bearings and get lost (we think we are our space suits); forgetting that we are divine spiritual beings inhabiting our physical bodies, traveling within form as a curriculum for awakening. We get sucked into the world of form itself and get turned out of our essence, as we believe we are limited and flawed human beings who are trying to find divinity, perfection and what we consider happiness somewhere "out there" in the world of form. The more we become a part of the consensus reality of the world of form the more we go to sleep and forget our true divine nature, and our connection to "all that is."

PHASE TWO is the layering process where we "learn" negative belief systems, fears, anxieties, and frameworks of limitations. We act them out and create affirming experiences that further re-inforce us in this cycle of negativity. Since the world of form is very sticky and dense, we become attached to it and identify with it. We think that our personal movies and everyday dramas are real. Our costume of physicality becomes so tight fitting, we think we are it. Our lens of perception is focused solely on the external material world. We believe we are both our physical body and our thoughts, emotions, and personality. We forget the multilevels of our existence and the multiple aspects of ourselves. We don't remember the true purpose of our journey. We suffer from a massive case of amnesia and we fall down, deep down into the webs and shadows of our unfolding dramas.

PHASE THREE is the unlayering process where we begin to discreate the "learnings" that have brought us to this predicament. We begin to come up for air heading toward the light. As we peel away the constricting layers of negativity our costume again fits much better and we can hear and see more distinctly. We start to open and awaken out of the separateness in which we had been imprisoned. We start to turn our physical world *outside-in* so that we can remember our connection with the greater part of our being; the totality of the Universe out of which our unique separateness (physical human experience) was created. We remember that we can honor our physical incarnation, while at the same time detaching from it and sitting quietly in the moment to moment choiceless awareness that lies behind it all. This process extricates us from attachment to identifying with our own dramas, so that we can play effectively in the world of form but at the same time still expand our perception to recognize the underlying greater part of our identity that lies beyond form, and out of which the forms were all created in the first place.

This is the concept of "completing the circle," experienced in Phases Three and Four where we start to awaken out of our unique physical dramas, extricate ourselves from our attachment and identification with them, and see the stuff of the world as being only partially real. We also use it as our vehicle to reconnect with the greater part of our being that is not limited by form. As we go from our unique separateness into the allness that lies behind it, we use our everyday world of experience to help take us out. Then as we sit in the oneness we so appreciate the beauty and richness of the world of forms, we come back down for more. We go down from our totality into our separateness. Being aware of the oneness that lies behind it all, we then can play and dance in the physical world of duality, honoring our unique separateness. And as we experience our separateness it constantly feeds back into the greater choiceless awareness—the unity that lies behind it, because we use it as the vehicle for this ride back up. This is a very fluid process like a cosmic elevator, going down and up, in and out, back and forth. We go down and lose our breath and then we come back up for air. The process is repeated over and over again. Our game of life is lived in the creative tension between the two aspects, and our *freedom* is to delight in this incarnation and use it as our vehicle to get back to our oneness. In this way we can live totally *in* the world but not be (sucked into all our dramas) *of* the world.

In PHASE FOUR we have now been able to reach a point of enlightened emptiness by relieving ourselves of the constricting layers of our negative programming. We have discreated identification with and attachment to our costume of physical form and we are ready to recreate for ourselves new layers of positive belief systems, insight, and true wisdom. Just as the Phase Two garment was ill fitting, dense and heavy, our new costume can be light, well fitting and porous so that we carry "a finely dressed conscious awareness." The porousness of our covering garment allows us to play fully in our world of separate forms without becoming stuck in it. It allows us to still see the reflected light of our underlying oneness which constitutes the greater part of our identity beyond the forms which it created. In this way we can more consistently focus into the perfection of the game of life where we use the dramas of our unique separateness, *which we fully experience*, as the vehicles to free ourselves from attachment/identification with them. It's like going from the crazy glue stickiness of layered Phase Two into the much lighter velcro connection of unlayered Phase Three.

We use all the experiences we have had in traveling through the prior phases to maximize the productivity and enjoyment of this phase. We understand and appreciate the multilevels and multilayers of ourselves and the world of consciousness which we inhabit. We understand the process of layering and unlayering; the fluid nature of completing the circle and its inherent remembering and forgetting, up and down, back and forthness. We cross the intellectual bridge of understanding gained through this knowledge, using an *outside-in* approach to more skillfully create for ourselves changed belief systems that produce affirming experiences to provide us with verification of our new levels of perception. We progress by taking the selves of us which we have changed for the better out into the market place of community so that we can also change our world for the better. We do this through a combination of helping and serving, compassionate action, and letting the light of our elevated consciousness attract, instruct and provide a mirror for others. As we continue to quiet our minds, open our hearts, and see the greater part of our being that is behind the whole dance of forms—we eventually assume the role of a modern day bodhissatva, a perfected enlightened being whose sole purpose in remaining on the earth plane of forms is to assist other beings in completing their journey of awakening.

We no longer waste energy trying to find our way to God because we realize we are already divine, and we have retraced our steps back to our forgotten divinity. It rests with us in the delicate balance we maintain between our veils of humanity and divinity. We have pierced the illusion of separation between the two so that there is a merging and interaction of the flowing energies between them. As our ego boundaries soften, we gain a stronger sense of our true greater identity.

Also at this level and in our journey to it we have been discreating the karmic effects of past life journeys! Our acquiring of wisdom erases karmic balances and the conducting of our lives on the earth plane with a greater sense of gentleness, kindness and grace supersedes the prior karma. Thus the work of our soul essence is further facilitated and the fulfilling of the cosmic blue print assured.

Our "work" at the physical plane level is also much more enjoyable. We don't need to receive such intense or painful learning experiences to gain our attention. We are already attentive and therefore our learning experiences become increasingly gentle, more positive and joyful.

In PHASE FIVE our vacation trip to the earth plane ends because we have completed the lessons of learning available in this particular lifetime. We shed our space suit, our sophisticated costume of form which we inhabited (we drop our body in the words of the Hindu's). It has served us well and lasted as long as possible. Ultimately it has worn out or been ripped or torn along the way, so that it now ceases to function. We leave the physicalness and density of the earth plane by raising our vibrational level to exit physical form so that we can return to our "home"—the macrocosmic realm of universal energy awareness. This "death" on the physical plane is a rebirth of the soul essence into the plane of "all that is" so that it can review its trip, the lessons learned, undergo additional curriculums at the higher celestial levels, plan its new blueprint for the next visit to the earth plane, and then start the five phased process all over again.

PART I
WHO/WHAT ARE WE AND WHERE DID WE COME FROM?

Placed on this isthmus of a middle state
A being darkly wise and rudely great
Created half to rise and half to fall
Great lord of all things, yet prey to all.
Sole judge of truth in endless error hurled,
The glory, jest and riddle of the world.
   —*ALEXANDER POPE (Essay on Man)*

Man is a rope stretched between the animal
and the superman—a rope over an abyss.
                —*FRIEDRICH NIETZSCHE*
                 (Thus Spake Zarathustra)

Man supposes that he directs his life and
governs his actions, when his existence is
irretrievably under the control of destiny.
                —*GOETHE*

1

PART I
WHO/WHAT ARE WE AND WHERE DID WE COME FROM?

PHILOSOPHERS ALL

Webster's dictionary defines a philosophy of life as "an overall vision of or attitude toward life and the purpose of life." It defines philosophy as the following:

a) Pursuit of wisdom.
b) A search for a general understanding of values and realities by chiefly speculative rather than observational means.
c) An analysis of the grounds of and concepts expressing fundamental belief.

And it further defines a philosopher as "one who seeks wisdom or enlightenment." But aren't we all philosophers of a sort, searching for our own truth and wisdom? One of the common characteristics of philosophers, going down through the alphabet of the ages of $\underline{A}$ristotle through $\underline{Z}$en masters seems to be this process of an inquiring mind that asks the questions and seeks answers directed to the larger issues of existence. We everyday philosophers wonder about certain important questions of life. They seem to have a natural progression, where the questioning process must take place at certain levels before other succeeding levels of inquiry are suggested or can be adequately explored.

This seems to be a never ending quest. However, the process of searching is essential since, paradoxically, the goal of our journey

3

is the journey itself; it has no ultimate destination other than to become aware of these concepts.

When I was a young boy I would eagerly look forward to Saturday. I could stay up late on Friday night, sleep late on Saturday morning and as I luxuriated in my bed, safe and warm beneath the covers, I would turn on the radio to listen to Big John and Sparky gleefully telling one and all "There's no school today." That program was followed by a show called "Let's Pretend." As I would listen to it, transported on flights of fantasy, my imagination world take me with the radio players to exotic destinations and involve me in heroic adventures far beyond my everyday musings. During the hour or two that I was living fully in these worlds of imagination, everything seemed so real and believable. I would even experience the physical sensations of sight and sound and smell and taste, and the emotional feeling responses of exhilaration and fear, joy and sadness. There was no differentiation between these two worlds.

I would also be reminded at these times of the equally graphic dreaming journeys that I would take and periodically remember from the sleep state. It seemed there were no clear boundary lines between waking reality, daydreams, flights of fantasy on the wings of imagination, and the drama-journeys of dreams. It wasn't until many years later when I began to question the validity of many of my assumptions and belief systems concerning consensus reality that I began to explore the possibility that there were "real" realms of existence beyond the material world of the physical senses, and that these invisible realities not only were equally valid but also existed and interacted simultaneously. This suggested many other possibilities about the nature of one's existence and further intensified the questioning process, that seems mandatory to all of us once a sufficient number of years and experiences tick their way off the clock of life. This questioning process which is encoded in all thinking human beings, is the way in which most of us give meaning, direction and purpose to our lives. It opens us to consider possibilities of identity and existence beyond our physical senses, our thinking mind, and the world it appears to create. What if our journey of awakening could be perceived in slow motion, so that we could be aware of the definitive phases that are the mile markers of our trip?

I'm going to ask you to spend one of those old Saturday mornings with me laying in bed, listening to the radio. Let's suspend our existing belief systems whatever they may be; step back for a moment from our consensus ideas of reality; loosen our belts, kick off our shoes, take a deep breath and let's consider the possibility that—

BIRTHING

Once upon a time; long, long, ago—the Universe was very, very quiet. There were no oceans or mountains; no plants, birds, or animals; no sights, sounds or smells; and no human beings. There was nothing except the deafening silence of the pulsating energy that constituted the open spaciousness of the Universe. It was an eternal energy force with rhythmical ebbs and flows, coming and going, and rising and passing away—moving ever moving in intricate patterns. It was circles of energy rippling in and out, out and in, changeless and yet ever changing. It was a kaleidoscope of pure consciousness that constituted "all that is."

Gradually, the universal energy essence began to create a world of forms by investing them with minute sparks of itself. The world became filled with billions and billions of things; living—breathing organisms of pure consciousness. There were innumerable colors and shapes and sights and sounds far beyond one's ability to describe in words or imagine in thought. The world became an incredibly diverse home for myriad physical forms, all apparently separate and distinct in appearance and function; yet all intimately related and connected as a spark of energy from the universal source of being. These physical forms were composed of incalculable combinations of animal, vegetable and mineral matter produced in diverse likenesses which spanned the gamut from physical terrains to bodies of water to plants, animals and human beings. Land and sky and water and creatures were all interacting in an exquisite dance of consciousness. The way of the Universe was the ebb and flow, rising and passing away of perpetual and constant change; yet with the appearance of solidity and structure by which one could perceive an ultimate order, harmony and purpose within the Natural Way of the Universal energy force.

In this fashion, the physical world became an amusement park for the senses. There was a continuous array of seeing and hearing and tasting and touching and smelling and thinking forms all created in an atmosphere of playful joy and wonder, and the human beings were the most interesting and unique of all. They were the most advanced creatures on earth and possessed a thinking mind and reasoning processes far beyond those of their fellow creatures. They were given the capacity to remember and be aware of the perfection of their divinity as a spark of energy from the Universal source, while at the same time they were living in the dense world of humanity and becoming involved in its dramas.

The greater part of our being is invisible, it is transcendent and exists beyond the limits of our ordinary experience of reality. That reality is limited to the physical world of our senses which we call "real" because we are able to experience it in a familiar and usual way. Since the invisible world of energy, consciousness and spirit can not be scientifically verified by the physical senses, we have a tendency to question its validity and we often label it as an illusion. But the world of physical form was created from this invisible realm. It exists as the true "home" of all of us and our physical world of forms is just a layer within that larger body of consciousness energy — within it, interacting with it, both separate from and connected to it, and most importantly—created by it.

These billions of sparks of universal energy essence exist at a soul level. At this level we souls are individual sparks of consciousness, parts of the universal energy essence. We are both separate and integrated within the larger portion of our being at the same time. At a point of our existence on this level, we decide to experience the three dimensional physical world of human beings for the purposes of learning and personal growth. While non-physical we exist in the planes of higher vibrational energy. In order to experience the physical realm we must lower our vibrational energies so that we attune with the density of the physical world. We draw up a sort of spiritual blueprint to determine what our growth lessons should be for this physical lifetime and the major learning experiences that we will encounter in order to be able to learn those lessons. This also includes the choice of our parents, the place and time of birth, our physical and personality characteristics, and any other related areas necessary to facilitate our completion of the blue print. This of course presupposes that we have experienced past lifetimes in the physical layer of existence.

This is not our first, nor will it be our last voyage to explore the earth plane. Someone once asked the Buddha how long we beings have been playing this game of life—this journey where the soul essence descends to experience the physical realm. He supposedly replied:

"Imagine that there is a mountain six miles high, six miles wide, and six miles deep. Every one hundred years a bird flies across the top of the mountain trailing a silk scarf in its beak. In the amount of time it would take that bird to wear down the mountain with the silk scarf, that's how long we've been doing this."

When our non-embodied energy essence has completed its past life adjustments we are ready for the new birth experience. These adjustments began when the soul essence was "reborn" back into the non-physical realm by the experience we commonly refer to as death. Then there is an extended review of the blueprint that had been created for that past lifetime, with special attention to the growth lessons and learning experiences that were successfully completed. There follows a period of further education and non-physical spiritual renewal at the end of which the new blueprint is created for the next incarnation. Finally the actual re-entry of the soul essence into the physical body of the infant birthed by the chosen parents occurs on the physical plane and the new life's journey begins.

THE LAYERING PROCESS

Dorothy Gilman describes one's awareness of the layering process in her book "A New Kind of Country," her personal odyssey of self-discovery:

"I felt the rigidities inside of me—the inhibitions and timidities and shoulds and oughts and musts and schedules and routines and tensions—as iron bands that encircle a barrel and hold it together by pressure."

A. ASTRAL LAYERING

The birthing process involves numerous personal choices which, in addition to a scripting of our major learning experiences, include one's parents, date-time-place of birth, gender, body type, physical characteristics, and emotional tendencies. Our chosen astrological configurations are the general indication of our personality—our predisposition to certain types of thought, feelings and behavior.

An early knowledge of our sun signs helps us to play the game of life with a full deck; knowing the subtle electromagnetic energy vibrations that stamp us with recognizable internal and external characteristics. They provide a predictable guide to our unique combination of human energy and emotional cycles, appearing as our traits or inherited tendencies.

My "sign" is the zone of the zodiac where the Sun was located at the exact time and place of my birth. The sun outlines my outer personality, the moon my inner emotions; and the additional eight planetary patterns further define my other behavioral characteristics—all with mathematical precision.

Awareness of this patterning is very helpful as a part of our growth curriculum in understanding the nature of our learning experiences, and how we will tend to react/respond to them. It also goes hand in hand with an understanding of our internal biological rhythms. How we function as human animals in our unique pattern of energy flow, digestion of food, elimination of wastes, sexual drives, physical attributes, and immunity to disease adds to our mental printout of self that helps us to understand many of the whys and wherefores of our lives.

B. INHERENT LAYERING

Now that we are beginning our Journey, it is appropriate that we look closely at what happens to us during our trip. First of all we are multi-layered beings and so is our world. We're like sparkling diamonds whose facets are numerous and overlapping, existing at ever subtler levels of consciousness.

We have our visible world of the physical senses where we feel comforted by familiar observational and behavioral realities. But there also is the invisible world of atoms and molecules and sub-atomic particles, all being a part of the universal and eternal energy essence. We have the illusory layers of sleep dreaming, day-dreaming, and altered states of consciousness induced naturally by meditation or intense personal experiences or artificially through chemicals. We are surrounded by psychic phenomena of all types, natures and descriptions. Communications from discarnate non-physical entities are channeled to us, we encounter increasing examples of telepathy, apparitions materialize and de-materialize and our extra-sensory perceptions heighten. Our inner intuitive senses are fine tuned and we experience increasing synchronicities and unexplainable abilities to foresee future events. In the physical realm of consensus reality the world of science and quantum physics daily expands its former conceptual boundaries to recognize and accept the validity of many of these aspects of the invisible world.

We are learning in truth and in fact that we and our world are not who or what we thought we were. Everything is multilayered and multileveled, each separate and distinct and at the same time all existing together simultaneously in a subtle and exquisite interrelationship separated only by the thinnest of invisible veils. All these levels of reality exist in the same relative time and space structures, acting and interacting like the ceaseless motion of the waves of the ocean of consciousness.

Everything is layered and pressed together like the finest plywood, so smooth and solid that only the minutest inspection and analysis suggests the differentiation of layers. That inspection and analysis process is what occurs when we begin our journey of awakening.

During our experience of this lifetime, our soul energy essence wears many different costumes as it bobs in and out of its layers of existence. And depending upon our personal level of understanding and awareness, we may think each of the separate costumes we wear are real and define the layered boundaries of who we are. For example we can look at our progression of personal awareness as it awakens through this fashion show of changing costumes. Ram Dass also suggests that our lens of perception flips through various filters:

1) The grossest layer is the one where I believe that I am my physical body. My view of the world becomes one of an external objective focus through my physical senses. Since I wholly identify with my spacesuit (physical vehicle), when it ceases to exist so do I and my perception of reality is limited to only this one lifetime. Naturally my experiences at this layer level will be directed primarily to physical gratification and recreational diversion.

2) In the next layer I identify with my thinking mind and become totally entrapped in my thoughts and sensations. I can't see that they are just mind-body processes that constantly arise, exist and pass away like all other kinds of phenomena of form.

3) I may next identify with my personality and see myself as the various psycho-social roles which I act out during this lifetime. This may include thinking that I am my profession as a doctor, lawyer, or indian chief, and I immerse in the deep and sticky melodrama that this layer level contains.

4) I may move out farther and believe that I am existing at the astral level so that I identify with my astrological houses. At this layer level there are only 12 costumes that I can wear corresponding to the 12 signs of the zodiac and my personality traits and emotional responses are governed by these configurations. Or I can identify with mythical archetypes such as hero, victim, good vs. evil, wise sage or courageous warrior.

5) As my awareness continues to open and expand, I may begin to appreciate that the apparent separateness that exists between myself and other physical beings is more an illusion than reality. When I look at another being I may be able to see the soul essence lying behind the eyes (windows of the soul) of that particular spacesuit that I'm observing. On this level we are all fellow soul energy essences, entities who are trying to figure out the nature of our Journey's in these costumes.

6) As I gain more awareness and understanding I may be able to end completely my identification with these changing physical forms. At this higher level I can understand that "there is only one of us." There are not different souls inhabiting different costumes, but rather it is all one universal energy that

is constantly changing and shifting in it's inherent light patterns of movement and creation.

7) If I am able to totally awaken, open to the greater part of my being, and pierce the illusion of separateness which had convinced me of the various other layers discussed—I may reach the Buddhist concept of the void where all layers melt away and no boundaries exist. There is only the unity of emptiness.

The great Indian saint Ramana Maharshi often responded to questions from his disciples about the nature of existence with the suggestion that they ask themselves "Who am I?" This questioning process served to peel away, layer by layer, their misconceptions of identity until what remained was the energy essence conscious awareness behind everything, noticing it all.

At one time or another we may occupy some or all of these levels of existence, but it's all a case of mistaken identity. We may look like the Vaudeville performer who is the fast change artist—wearing different costumes and playing separate roles as fast as the audience can follow, but we really are *all* the categories flowing into each other in a constant dance of consciousness.

C. Societal Layering

Multi-layered beings that we inherently are, we also externally layer ourselves on the physical level by becoming sucked into our gross physical world of the senses. We become so seduced by the ever changing landscape of sights, sounds and feelings that we identify with the soap operas and melodramas that we are living. This may be described as the *ontological* stages of our life awareness—relating to body, mind, and spirit:

A) From age 0 to 5 we essentially have a clean slate that we brought with us from the higher vibrations of the soul essence level. We are exemplified through the innocence, openness and spontaneity of young children. At this level there is still a remembrance of our divine origin and the nature of our perfection. We have basic needs for survival that are genetic in origin in order to assure that the species will continue. They relate to food, warmth, shelter, and the fundamental need for love and nurturance from other human beings. At this level we are more in our spiritual divinity than in our material humanity as we exist totally in our moment-to-moment choiceless awareness.

B) Rapidly, after age 5 and through adolescence, a veil of forgetfulness descends and any recollection we have had of the blueprint for our journey and our divine connection to our non-physical home is extinguished. Our "learning" process begins and with it our focus of perception becomes external and governed by our world *out there*. Our primary teachers are our parents who usually do not have much of an understanding or awareness of their own identity. They hardly understand what the game of life is, let alone trying to teach it to their child. This is where we start to experience and learn negative separating emotions such as fear, anger, loneliness, and anxiety which start to layer themselves around our heretofore divinely immaculate openness. We begin to believe that we are limited beings, somehow less than whole, and lacking certain essential ingredients for completion. These requirements are seemingly unobtainable and the world appears to be threatening and hostile. In order to satisfy our perceived needs, we must cultivate our ego perceptions that we are somebody who is very important, focus our attention outward to do battle in a frightening world, and periodically engage it and retreat from it.

C) The un-learning process continues and accelerates through adulthood as our focus of perception changes from our inner world of unitive connection with the Universe to our externally focused world of duality and ego separateness. We become totally seduced by the world of the senses. Our life becomes a mad-dash toward sensory stimulation, materialistic achievement and recreational escape from our everyday perceived problems. As we layer over more and more of our being we continue to shut down our intuitive feeling levels and vision of the greater part of our being. Our belief systems become more negative and fearful in accordance with the ever increasing layering. Our actual life experiences also change in the same manner. We create a physical reality which mirrors the negativity and fearfulness which permeates our beliefs, thoughts, and ideas. We become lost in this vicious cycle of negative self-fulfillment, trapped in self-created prisons where we view ourselves as limited, vulnerable and threatened. We desperately seek to escape by pursuing pleasure and avoiding pain. This external struggling only serves to pull the harmful layers of programming ever tighter around us. We constantly reinforce our negative belief systems that started the process in the first place. This is an insidious process where the more we focus on the exterior world of objects and sense pleasures for happiness, fulfillment or completion, the stronger are our expectations that our need satisfaction will come from that outer area of focus. When this does not occur we create additional layers of disillusionment and disappointment.

We become angry and cynical, full-time members of consensus reality. This in turn further strengthens and reinforces the negative emotional layers that are already in place. They are very sticky, like the strongest fly paper. As we continue to attract to ourselves negative experiences that naturally flow from our increasingly negative belief systems the layering continues and becomes more pervasive. Ultimately, our natural energy essence becomes constricted and blocked. We crystalize into a solid encrusted form that functions, if at all, like a robot automatically reacting to our programming of duality and negativity. We are in a state of non-awareness where we live our daily self-created heavy dramas of existence. We believe them to be very real. We are alive in the physiological sense but are dead to the joys of living a conscious life. Just as the initial phase of infancy was an age of playful *innocence*, we have now reached the *end of innocence* and time hangs heavy on our hands. We cannot breathe, and our suffering is intense. We have lost consciousness, and have fallen fast asleep.

UNLAYERING

Eventually, we all have a chance to regain our greater awareness and awaken. This usually does not occur however until we go through certain rites of passage where we experience our seemingly mandatory period of un-learning, layering in of negative belief systems, attraction to us of actualized negative experiences, increased layering, loss of vision and energy, helplessness and despair. It seems that our questioning process and our efforts to pull ourselves out of this quicksand do not adequately begin until we develop a strong standard of comparison. We must both understand intellectually and feel experientially that we have strayed off the correct path in our journey and it's not working for us. Then and only then do we seem to have sufficient motivation to look elsewhere. The external world of sense pleasures is and has been very distracting. It is hard for us to turn away from it while our lives seem to be working in a manner that is satisfactory.

Ultimately the maturing of our awareness that "there must be more to life than this" is completed, usually around mid-life at age 35-45. It often coincides with a life experience of extreme stress or unusual intensity such as serious illness, career crisis or the end of a love relationship which provides sufficient fire to start our questioning process and the melting away of our encrusted layers of consciousness.

It's as if we had armored ourselves with a huge steel door that became progressively more rusted and corroded as we travelled through the storms of life. At the point where it becomes totally sealed, we may experience "the dark night of the soul" when in the desperation of our pain and suffering we are presented with the opportunity to open up once again and experience the fullness of life. Eastern mystics have said, "At the darkest time of night, never are we nearer to the light." This is the beginning of the unlayering process and can be a wrenching experience as our door of consciousness is forced open.

As our unlayering process progresses, we encounter more people, places and things to provide learning experiences and opportunities for further inner growth. Although intense suffering may have been necessary at first to get our attention, soften our rusted doors of ego structure, and lessen our resistance — we now are opening. The more we do so the less pain and suffering is needed to move us forward. The lessons become less difficult and intense. They don't need to get our attention so dramatically because, in our opening and heightening of awareness, we are *already atten-tive*. More joy and less pain becomes our everyday experience of life.

We begin to discreate the negative programming that had formed the basis for our overlayering of ourselves. Similar to the experiences we had while unconscious and trapped within a negative belief system, we now become more positive in our thoughts and this outlook is confirmed by the actualizing of more positive everyday experiences. This in turn further reinforces the positive beliefs which creates more affirming experiences and so on.

In the layering process, ego was created as the set of thoughts and sensations that we use as perceptual filters to view our world. It is our conceptual structure of ourself and how we function in our unique Universe. Now that we are unlayering and shedding the veils that had kept us from our inherent awareness of the overall structure of ourselves and our Journey, we may be tempted to identify our ego as the villain and try to destroy it.

But the idea is not to obliterate the ego; rather it's just not to be trapped within its confines, to the exclusion of the "big picture" of life. Our ego is important. It enabled us to develop our basic sense of self, so that we were grounded enough to function on the physical plane. It serves as the daily launching pad for our Journey. We need enough of it to maintain our creative tension as we balance between the world of our humanity as we honor our incarnation, and rest calmly in the world of our divinity which lies behind it all.

RELAYERING

The process continues as we retrace our steps and not only shed the layers of negative programming that had previously inhibited us, but replace them with positive programming that creates a more fulfilling reality for us. This is not by any means a smooth and continuous process; rather just as we rode the roller coaster between pleasure and pain during the layering and unlayering processes we now ride it again through our numerous experiences of remembering and forgetting, as we gradually move away from our former constrictions into our expanding levels of awareness. We find that in this process of awakening, we shed the skins that had prevented us from fully experiencing a happy and fulfilling life, and become new, more sleek versions of us. We encounter times when we are fully awake and open to the greater part of our beings, and other times we doze and revert back to acting out our former negative belief system. However we notice that no matter how many naps we might take we never fully go back to sleep again. We accomplish this process by essentially turning our world *outside-in*, and starting over again in a much more aware and skillful way.

As we open to the greater part of ourselves the world in which we live doesn't change, nor do we change in outward physical appearance. However, the inner transition is profound, as we gain insight and awareness into what life is really all about. Our actual experiences of life begin to change for the better, as we do.

We ultimately reach a point in our quest where our waking moments are more frequent then our lapses into unconsciousness. As our personal healing progresses and we again welcome a sense of our wholeness, we experience a strong feeling that we not only are happy to be changing ourselves for the better, but we want to take our higher consciousness out into the market place of our everyday life and try to change our world for the better. The sense of helping and serving others becomes very important, both for personal fulfillment and in a more Universal sense of the oneness of existence.

At this layer of being we encounter *illuminated innocence* where we have unlayered sufficiently to regain our earlier vision of perfection—the love and lightness that is us. We become reacquainted with the perfection of our soul essence while at the same time bringing to it the wisdom we have gained through our life's journey thus far. The more we open and personify our heightened consciousness through kindness, caring and compassionate action, the more we experience the deeper realms of unconditional love which forms the basis for the Universe to which we are always connected. We now are able to play the game wisely and to see its many layered aspects, all interwoven in subtle and exquisite ways.

ACHIEVING BUDDHAHOOD

As we heighten our levels of awareness and attract more affirming positive experiences we become wiser in our journeys. This wisdom erases karmic balances that may have existed from prior lifetimes where we were less skillful. As we refocus on our connection with the Universal energy essence and reflect outward that heightened awareness in our everyday experiences, our lives become more filled with a sense of grace and an inner contentment. We are able to utilize our own free will and choice through our heightened and compassionate action in this lifetime, to create our own most desired realities and mitigate the negative aspects of past lifetimes.

The ultimate stage of life awareness is where we achieve a sort of divinity on earth. We are free, having no further attachments of our thinking mind that cause us to suffer. We no longer cling to what we have or grasp at what we don't have. We accept life as it comes, and also trust that the best will find us. We even accept the world of perpetual change and do not stand anywhere in it. We hold on to it tightly in moment to moment experience, but are also able to let it go lightly at any time. We are in the world but not of the world. We go through life perfectly balanced on the razor thin tightrope between our humanity and our divinity. By achieving this delicate balance we have reached enlightenment and we assume the role of bodhisattva, where our primary purpose in continued physical existence is to help to bring others to this level of freedom, as we continue to live conscious and caring lives.

This enlightened state is not an event to be achieved or an identity to be grasped. Rather, as Jack Kornfield says, "it is a deep opening to the ever-changing natural cycles of life; in us, around us and through us."

PART II
WHY ARE WE HERE?

To myself I seem to have been only like a boy
playing on the sea-shore, and diverting myself
in now and then finding a smoother pebble, or
a prettier shell than ordinary, whilst the great
ocean of truth lay all undiscovered before me.
                    —*SIR ISAAC NEWTON*

Conquer thyself. Till thou hast done this, thou
art but a slave for it is almost as well to be
subjected to another's appetite as to thine own.
                    —*ROBERT BURTON*
                    (Anatomy of Melancholy)

We are all born for love . . . It is the
principle of existence and its only end.
        —*BENJAMIN DISRAELI (Sybil)*

PART II
WHY ARE WE HERE?

THE DILEMMA OF PURPOSE

That same aspect of ourselves (that we explored in Part I) that seeks it's identity beyond the ego boundaries of the thinking mind, is equally questioning concerning the purpose of our existence. Once we have a better conceptual understanding of our multifaceted nature and the multileveled universe in which we exist, we are still incomplete, fragmented in understanding, until we connect both the event of this visit by our soul energy conscious awareness to the physical earth plane with the reason for the trip. So we ask, "What is it I'm supposed to be doing here?"

Understanding and appreciating the stages of the journey, the processes of layered learning, unlayered discreation, and positive relayering all point to method behind this apparent madness. As singular as is our divine plan relating to our real identity, nowhere is it more exquisite than in an examination of the basic objectives for our existence that lie behind it all.

The mystical texts, scriptures, and spiritual literature down through the ages talk about man's quest being that of "coming home to God" or reclaiming one's inherent divine nature. This is the idea of a point in the time sequence of existence when man's seemingly endless succession of troubles are over, he stops identifying with his worldly dramas, he opens to the greater part of himself, and he returns to the bosom of his Creator. Whether couched in terms of religion, spirituality, mysticism or transpersonal psychology, the basic thrust of all this is to find a path away from our acquired duality of separateness back into unity consciousness where there is only one of us.

THE DIVINE PLAN

We can continue the New Age catechism of "Who am I?" that began in Part I with our Part II question "Why am I here?"

1) I am a soul made manifest in matter who becomes lost in the manifestation, and who then reawakens to my divine soul nature.
2) I am pure energy conscious awareness, a spark of which has manifested in the earth plane of form, for the purpose of experiencing it and learning to fully understand all aspects of it.
3) I am soul awareness, visiting this physical plane and getting lost in my dramas of incarnation, for the purpose of awakening through this experience, and learning certain karmic lessons.
4) I am a part of the vast universal energy gestalt, that chose to take a physical birth and totally identify with the earth plane of physical form, to learn to reawaken to my divine nature while at the same time honoring my human incarnation.

This "I am" litany can continue indefinitely but behind it all is the basic assumption that there is a plan for our existence and this is not a random universe. If random, our lives would be purposeless, there would be no meaning to it all, and we would be like leaves blowing in the breeze of change with no pattern, objective or cause and effect relationships. However we know this is not the case; our head and heart tells us otherwise. Our physical senses verify clues of the inherent harmony of things as we view the dazzling array of forms in nature. There is total diversity and at the same time a basic underlying structure of connection and interrelationship that translates into a Universal intelligence behind the plan and a sense of order and justice which prevails. Intuitively we also have a sense of meaning and purpose. Our internal questioning process, which is an integral part of this Universal creative thrust, keeps pushing us in the direction of finding answers to our questions in experiences that affirm the cycles of growth.

As we expand our awareness we are able to see the interconnections and multiple levels of identity and existence. The totality of it all includes the familiar physical world of forms on the earth plane, but only as a part of a greater whole. We see that everything is mutually inclusive, *not either—or*! Our ego thinking mind is by it's very nature mutually exclusive, moving back and forth between polar extremes. Since the rational mind is the basis of our western society, our lives are filled with this process of always judging, evaluating, criticizing and comparing separate phenomena. There is always a barrier between, rather than a merging of the all. The statement, "When we are able to go out of our mind, we can regain our sanity."—makes sense at this level of understanding.

This expanded vantage point includes all experiences and all realms of existence; our familiar and visible world of form and the more ethereal invisible world of moment to moment choiceless awareness. This brings to us an understanding and appreciation that we are not here by accident, the living of our lives is not some type of cosmic error, and everything is connected in a cause and effect relationship.

BALANCING OUR KARMA

The greater part of our being, our conscious awareness that lies beyond our form, our soul energy essence—is also known as our higher self. This higher self chooses to take a physical birth and visit this earth plane of form in order to have certain experiences which only the physical plane can offer. These experiences are to balance and complete certain spiritual assignments of learning and growth. This greater part of ourselves, our soul essence which is embodied in physical form, goes to school in the earth plane and uses its intensity of experience to learn important lessons of growth and awareness. These lessons are not predestined but rather are predetermined at the soul level in the form of our spiritual blueprint. The spirit essence, wearing its physical costume, has free will and choice as to the manner and method of learning the lessons. How we react or respond to the experiences which are our curriculum for learning determines what we characterize as happiness or suffering.

This is why it is said that we are only in our bodies for experience. We are here to learn first for ourselves, and then to be able to radiate those self-learnings out into our world of experience and interpersonal relationship. This is the process of integrating our spirit essence, through form, so that the broad search in any lifetime is to come to know the greater part of our connection to universal energy consciousness while at the same time honoring our humanity in the classroom of earth. Why else would the soul essence force itself into the tight costume of the physical body and the world of forms?

This constant striving to reunite with the divinity that is our essence; to remember forgotten aspects and reconnect with our primordial experience as one vast energy essence/conscious awareness is why we keep doing this over and over again. We use the density of the earth plane as the rough sandpaper, the grist for the mill in our learnings.

So we have an overall universal learning experience for returning to our divine nature, and the specific learnings for each lifetime that comprise our spiritual blueprint. Reincarnation and karma go hand in hand to explain what we are doing here, why we keep coming back to do it, and also the justification for the apparent inequalities that exist in our lives and our world.

Karma has been described as cause and effect, "as you reap so shall you sow," a system of accumulating karmic debts. By our actions we pay off those debts to balance out the cosmic books, if we are living conscious lives of love, kindness and compassion. The golden rule, "As we do unto others so shall they do unto us," is really nothing more than a statement of the principle of karma.

If we pick up a stone and throw it into a still forest pool, the thrower of the stone is the cause and the ripples are the effect. The action has created disharmony in the pond and the ripples flow out and back until finally the pond returns to its original state of balance. In a similar manner all our thoughts, words and deeds have an effect upon the balanced harmony of the Universe. Everything we do in this fashion creates vibrations that flow out and back through all lifetimes, possibly disturbing the balanced harmony and tranquility of the Universe. If that is the case these karmic consequences must eventually be balanced and the "pool" of universal conscious awareness must be returned to its original harmonious state. Just like the flowing and interweaving energy of the Universe, karma is constantly in motion acting on every level of our existence. As Dick Sutphen suggests, "everything we think, feel, say or do has karmic effects, as well as the motive, intentions, and desires behind these acts."

Thus if we think, feel or act with negativity we generate a certain amount of negative karma that will be experienced and manifested in our lives. If the motive, intention or desire behind our thoughts feelings and deeds is positive, conscious or beneficial then we generate positive karma. It's a system of debits and credits which accrue to our personal ledger of cosmic accountabilities. If we help and serve out of true kindness, caring and compassion, we generate positive karma. If we help in order to impress others or boost our ego about what a good person we are, we really are generating negative karma. More important then the content of *what* we do is the *why* of doing it.

The karmic balances are constantly being totaled as we exercise our choices of experience during our lifetime. The actualized experience that we are presented with, as being positive or negative, is really a creation of this balancing process as our karma runs off. Therefore every aspect of our life is the result of this continual process of cause and effect; including our physical body, our state of health, our prosperity or lack of it, our experience of relationships, and everything else. We are constantly being presented with learning experiences and opportunities for growth that are manifested from this cause and effect process. Our karma acts like a psychic DNA that manifests in our cosmic blueprint of potential dramas.

Therefore we are totally self-responsible for everything that has ever happened to us. This is a heavy statement to consider but it is nevertheless true. There is no one to blame for anything that has ever happened, there are no victims, and there are no accidents. Every experience has been created by us, through our thoughts, feelings and actions, steered by our motives, and the intentions and desires behind them. We are constantly creating our own reality in this fashion. We are really expressing our divinity by being the Creator of our moment to moment experience.

The karmic flow presents us with continuous opportunities to

directly experience the consequences of our actions. My acts or
failures to act create my karma. If I buy into your dramas, I assume your karma. Since the fastest way to learn in this school we
are attending on the earth plane is to directly experience the lessons—the more we tackle them head on and go through them, the
quicker we will graduate. Since we always have free will and
choice as to whether we will deal with such situations and the
manner in which we do so, again we are totally self-responsible for
not only our experiences but also the pace of our personal growth.

In addition to the karmic consequences of our everyday actions
we bring with us into this lifetime the balances of stored up karma
from past lifetimes, since not everything can be balanced in one
lifetime. This birth karma, as opposed to our day to day karma, is
usually a part of the game plan blueprints that we create on the
soul level, as we design the major learning experiences that we
will encounter on our particular visit to the earth plane. The results of our experiences in past lifetimes and the karmic balances
which were generated helps to explain the seeming inequality in
the world around us. Obviously the life experiences being encountered by different beings are a direct manifestation of their past
karma.

For example, while between lifetimes on the soul level we may
decide to create learning experiences in the next lifetime in the
area of stormy and negative relationships with another being
with whom we have shared prior lifetimes. We create a script for
this forthcoming lifetime—we will ultimately meet each other after we have both had a series of failed relationships. We will have a
relationship initially characterized by the similar intense patterns of our past lifetimes, but then we will awaken from these
recurring undesirable patterns, expand our awareness, learn our
lessons, and complete this growth experience with a committed
relationship of mutual love and respect. We then initial the script,
find parents with an appropriate karmic configuration that fits
our need, leave the "other side" and go down to the lower densities
of the earth plane by taking birth, and begin our movie of learning
experiences. At the conclusion of the lifetime, we are reborn back
into the higher plane which we call the soul level. As we replay our
movie just concluded, we see that we have balanced the karma in
this relationship issue, will not have to repeat it, and can now

move on and create a new script for another learning on our next visit down to the earth plane.

In the interplay of these karmic causes and effects, just as we have free will and choice as to how we deal on the earth plane with predestined learning experiences, so also do we have the opportunity to erase karmic balances by how we live our lives in this lifetime. If we live a life filled with expanding awareness, kindness, compassion and loving—we can run off the balances of negative karma that had existed from prior lifetimes. We also can reduce the amount of undesirable future karma that we may have awaiting us as part of the specific predetermined learning experiences that are our soul scripts for this and future lifetimes. This mitigation of karmic burdens is another aspect of our self-responsibility and a manifestation of our divinity where we can create our own experiences of a desired reality.

An integral part of using our visit to the earth plane of existence (our incarnation) to evolve spiritually, is to view its purpose not only from above in the cosmic realms of karma and reincarnation, but rather from below on the earth plane of existence in the context of our everyday lives.

CALMING OUR MIND—
The Illusion of Identification with Our Dramas

In a very practical sense, our purpose here is to alleviate suffering, both ours and our fellow beings. This is the basic philosophy of the Buddha in his four noble truths examined in detail in Part IV. This suffering is both the cause and effect of our sense of separateness, isolation and helplessness. It all has a further manifestation in our world which seems out of control and on the verge of planetary extinction. Reducing our suffering helps us individually and collectively.

In order to alleviate our self-created suffering we must detach from the clinging nature of our minds, we must disidentify with our personal dramas, we must find our way out of being lost in our personal story lines. We know that while we live in these sticky attachments of our mind we create a life experience which is consistent with them. On the expanded level of our awareness we know everything is just impartial phenomena. It is all just mind-body processes which are neither good nor bad, positive nor negative, neither this nor that. It just all is. However our ego thinking mind is a conceptual mind which continuously creates our world for us and our experience of living in that world by going out and grabbing at the phenomena and giving it meaning in its own terms. Since inherently it is a mechanism that evaluates everything as either being a threat or non-threat to its existence and it labels and judges accordingly, our experience of life at this level of form is extreme agitation and we suffer.

If we can quiet the noisy chatter of our mind we can find that part of our awareness that is not entrapped in thought and drama. We can go one step beyond the dramas, stand back from them, and rest in that space of conscious knowingness and expanded awareness. When we can thus disidentify we are in a quiet present state of moment to moment choiceless awareness from which thinking happens but in which we do not identify and become lost in our thoughts. This is the spaciousness that is necessary to alleviate our suffering, and which allows us to move deeper into living a conscious life. This is the place where we are aware, and watch ourselves creating and participating in our personal dramas as a necessary experience to learn our lessons on the earth plane (thus honoring our human incarnation). But at the same time we can maintain our awareness of the connection to the greater part of our beingness which birthed the human part in the first place. We are trying to integrate both our overlying divinity with our underlying humanity, balancing them both in a positive and constructive manner. We live our lives in the tension between these alternate pulsations of energy—Satori or soap opera. When we suffer, we lose sight of this eternal connection, and become lost in one of the extremes. We are detoured off the main road in our journey of learning. But in reality we never lost *it*.

It is the greater part of our being, our divinity, the pure energy essence choiceless awareness that is our true being. *It* is beyond the physical body and its senses, beyond the personality and feelings and emotions, beyond the clinging nature of our thinking mind, and beyond the walled boundaries of our ego. *It* is beyond anything and everything that is limited by form. *It* is the place from which all form, including the little us that visits the earth plane, is created. *It* is a deep quiet place of all the wisdom of the ages and the totality of experience gained in all our prior life-times. *It* is the elevated perspective of the whole game of life beyond time or space. *It* is unlimited, eternal, permanent and unchanging, impersonal and connected to all that is. From *It* is created everything else and *It* is the source of the Universe.

In order to calm my mind I have to locate that detached witness —impartial observer part of myself that is behind my dramas of physical form noticing it all happening, and sit there for awhile. I want to feel what it's like to experience and appreciate both levels coexisting simultaneously. Then from this elevated vantage point I can notice not only that my ego thinking mind is clinging and creating suffering, but where the clinging is occurring. When I can see where I am identifying with my personal dramas and where I am attached and holding on to them, I recognize that those spaces of holding are very tight and constricted. In them there is no awareness of the greater part of my being that is watching the whole show.

Then after I sit on this ladder for awhile, relax, and watch the whole show—I can practice descending with the necessary spaciousness to go behind the dramas. I can start to see that I am not my soap operas, it is just neutral phenomena and mind—body processes that my ego thinking mind has convinced me is real and serious. These are thought forms that I have identified with as reality. As I quiet my mind, there is an automatic lifting and lightening process where the stickiness and clinging and holding of my mind releases and I am able to rest in this spaciousness where I can see and acknowledge the dramas but know I have a choice not to identify with them. The more I do not identify with them and disconnect from them, the more the clinging of my mind releases and the suffering experience diminishes.

In this fashion I learn that I am not the actor, I am just the creator and observer of the actions. I can regain my sense of pure conscious awareness by getting behind this mind stuff. I am really trapped in prisons of my own mind, rather than any exterior experience. The clinging and attachment and identification with my mind creates the cell which I voluntarily occupy until I break out or am paroled by awakening out of identification with all this thinking mind ego drama of the world of form. Until I do so I'll continue to do time in this prison that I have been creating.

Once the thinking mind changes its perception from the idea

that the personal drama is real to consider the possibility that it is just a thought form provided as the curriculum to awaken from consensus trance—we gain the larger picture of existence. Here our suffering is just another of the transitory and fleeting aspects of form.

Chogyam Trungpa Rimpoche has said that "Enlightenment is the ego's ultimate disappointment." The ego is tenacious and it releases its hold very grudgingly. It's not that we want to kill the ego, because we need a certain basic amount of ego thinking mind consciousness in order to function effectively on the earth plane. Rather we just want to modify its exclusive status as the dictator and sole significant other in our lives. In actuality, we want to take another lover that is the greater part of ourselves, which created the ego thinking mind forms in the first place.

As the ego's role changes and we become more masters than servants, causes rather than effects, we see that while we were identified with the world of form we had abdicated our free will and choice. We assumed the role of a victim where everything "out there" was doing it to us as we were wrapped up in our dramas. To the extent that we now stand behind the forms we see that we are the larger awareness that is the creator, rather than the created. Rather than being buffeted about like a leaf blowing in the breeze, we see the perfect flowing way of things where we are the wind that propels the leaves.

As our mind quiets we also gain better perspective on the true nature of our quest. We are better able to absorb and understand the divine plan with all its ramifications. We recognize that we're not here to find our way to God—we already are divine. We don't view ourselves as nothing and God as everything. Rather, God is a part of our basic nature which we have forgotten and lost. As Swami Muktananda said, "God dwells within us, as us—om namah shivaya." We just need to retrace our steps by moving *outside-in*, back to the divinity that is the greater part of our conscious awareness. When we do that we extricate ourselves from our dramas. It's like army maneuvers where peace has suddenly been declared—"About face soldier! Let's make love, not war!"

Quieting of the mind is best achieved through meditative and contemplative activities by virtue of which we cultivate a calm noticing of phenomena and an awareness of that greater part of our being quietly watching all the drama. In this place of stillness, we can know our true wisdom. We can hear clearly so that our actions come out of the wisdom of silence rather than the confused chatter of the ego thinking mind. Rather than going through life as a reactive robot, mechanically acting out the roles in my dramas which are assigned by my mind, I can awaken from that type of consensus reality and responsively live a more conscious and meaningful life. I don't have to act upon the thousand and one wandering thoughts of my busy mind. I try to just notice them as they rise and fall, withholding my habitual inclination to react.

The concept of meditation can be confusing. Numerous books and treatises have been written about it. The meditative space of quietness and inner knowing which accesses the greater part of our being can be reached in any number of ways. This can include but is not limited to formal meditative practices such as zazen sitting meditation, insight meditation of following the breath, dynamic moving meditations, kundalini energy tantric practices, various forms of yoga, visualization techniques, sound and light technology to alter states of consciousness, mechanistic brain hemispheric synchronization processes, and related relaxation techniques. In reality, the times when we are totally absorbed in any activity so that we have no conception of time or space, which could be reading or writing or dancing or listening to music or admiring nature—are all forms of meditation. The labeling that we do as to whether or not a particular type is better or worse or more or less efficient is simply another of the thought form judging games of our ego mind. Meditation is not a form of anything, it is a space that we open to which is beyond form. The word "meditation" more properly describes the state of quiet mind which is the goal, rather than the method used to achieve it.

OPENING OUR HEART—The Illusion of Separateness

As our conscious awareness passes through this lifetime on its visit to the earth plane occupying a human form it is struck by the tremendous amount of suffering that exists. One of our defensive reactions is to close down our heart and try to push away the suffering by pretending it doesn't exist. This is done because it is so intense we are afraid that it will burn us up if we open to it. We are afraid that our heart will break if we allow the suffering in, so we build our illusion of separateness through the walls of ego and make the whole world "out there" into them (our adversary) in order to protect ourselves. Unfortunately by doing this we are cutting ourselves off from the juice of life which nourishes us. Our heart is like an umbilical cord to the nourishing joy of living and when we close it down we starve.

On the higher level of awareness our intuitive heart is always connected to all of the multilevels of our reality. It integrates the realms of divinity and humanity and accepts everything as the stuff of life. But our emotional romantic heart is connected to our ego thinking mind so it lives in the limited world of form. The suffering that is the result of karmic lessons, clinging of mind, and separateness causes it great pain. This ego heart protects itself from claims, demands, or perceived threats of others and generates it's own type of additional suffering as we isolate ourselves from our fellow beings. If we remove our nurturing we suffer the effects of spiritual malnutrition.

When we go up into the greater part of our beingness we see the basic oneness of everything. We are all one, just dancing as the many. When we go down into the separateness of our world of form we live in duality and our world of me and mine. The game of course, like all the games within games we have been talking about, is to integrate both in our moment to moment experience. This again involves the paradox of our being divine and human at the same time. We do not live in a mutually exclusive world of polarities where everything is either-or black or white. Rather our true world is mutually inclusive where everything is an integral part. We are both human and divine. Life is perfect and it is ghastly. I'm happy and I'm sad. Everything is shaded in terms of gray, rather than black or white as we live in these constantly merging and interconnecting worlds at the same time. Both our physical world of form which is visible to the senses and our greater spiritual world which is invisible and metaphysical exist and interconnect at the same moment-ness. We are constantly balancing on a razor thin tightrope between these worlds of unity and duality, inclusiveness and separateness, the clinging mind and the spacious awareness—all existing in a continuing panorama of life.

If we have the patience, awareness, and determination to eventually embrace unto ourselves life's suffering, recognizing it as an integral part of the journey, we can also become aware that it is forcing our ultimate awakening. It is the stuff of our learning. It gets our attention so that we are more attentive. It cracks open our ego heart which is caught in the world of form. As that heart breaks we open into our intuitive heart which lies behind it, beyond the world of form, with no limiting boundaries. Everything is included in our greater heart, even the suffering, because the painful experiences act like rough sandpaper that keeps smoothing away the encrusted attachments of our minds, freeing us from being lost in our dramas of life. They burn off the hooks that catch us, that we grab and hold onto. When there is no place left to hold, we also have no place to hide, and our learning is accelerated into areas of open spaciousness.

Bo Lozoff talks about not shielding ourselves from suffering. He suggests allowing our heart to break open again and again, so that we can expand it continually to hold more and more love. This is honoring our humanity at its highest levels. Behind all the drama there is really only one of us. If I injure you, I'm hurting myself, so I must strive for unity—not duality. And the vehicle that gets me there is love.

TAKING THE CURRICULUM—
The Illusion of Pushing Away the World

The earth plane is such an exquisite training ground for us. It is so easy to get lost in our personal movies and soap operas and think that they are real. It's so gooey and thick with drama; chaotic and at the same time very exciting to be lost in. The world of form is also very visible to our physical senses. We can constantly verify its apparent reality rather than the invisible and quiet world of the One from which we came and from which these forms were created. There is such a thin etherial veil between the two that we believe the alternate model of reality that is created by our thinking mind. It throws out the nets of illusion that create the models of our world and our required roles in it. When we move out from the mind we can see the true picture. The mind is unable to be an impartial observer behind the forms, because they are all a product of the mind. It can't stand back from form, because it is an aspect of form itself.

But from the perspective of the greater part of our awareness that is not entrapped in thought, we can see that our everyday experiences in this world of form provide the curriculum that gives us the opportunity to awaken out of it. At the same time we become lost in our dramas in the world of form we can utilize them as our vehicle to awaken out of it into that greater part of us that is beyond form. Thus the forms themselves can help us to awaken out of our attachment to them and identification with them. This includes our body, our personality, our thoughts and feelings, and our ego thinking mind because they are all a part of that world. That part of me beyond form (my energy essence conscious awareness) is attending school and using the everyday experiences of the world of form as it's curriculum to graduate. My body, personality, feelings, emotions, sensations and ego thinking mind does not take the curriculum—rather it *is* the curriculum.

Therefore I'm not really who I thought I was when I started this journey in Part I and as Ram Dass says, "went into somebody training." At that time I thought I was my body or my personality or my emotions—feelings or my ego thinking mind. I thought that was reality. Now I understand it's all part of a larger all-encompassing reality.

I (the greater part of me beyond form) have to get lost in my dramas and entrapped in form before *I* can be extricated from it. I have to go to sleep before *I* can awaken. *I* have to become somebody before *I* can become nobody—that's the process!

As I fulfill my soul's purpose by using my daily experiences as the curriculum for awakening, I am able to open out of the illusion of duality which is embodied by my human form. I can experience the balance where I'm connected to the totality of everything in the Universe and at the same time I honor the unique separateness of my human incarnation that came out of the One.

I have to honor my incarnation and take my mind—body stuff seriously if I want to function on this plane. I have to know my zip code, live in the marketplace, make my mortgage payments, drive in heavy traffic. In doing so I will frequently get lost in the dramas. I have to get caught in them sometimes since I'm taking the curriculum and these aspects of the world of form are the curriculum. However, at the same time as I play effectively in the world of form

I can still honor my connection to that greater part of myself beyond it. In that fashion I can be in the world, playing the game
with intensity and effectiveness, but at the same time not be of the
world where I get totally lost in the dramas. From this perspective
I can see the veils which divide the multilevels of the Universe. I
can honor them all and do my dance of life in the creative tension
that exists between them. I can honor the earth plane as a beautiful form but not get lost during my visit.

Yogis and mystics may gain great powers through their withdrawal from the world, monastic practices, and undergoing of
rigid austerities—but if in doing so they are avoiding the curriculum of experiencing their everyday life their learning is incomplete. It's easier to leave our busy world and live in a cave then ride
the daily roller-coaster. Our maximum growth, however, comes
from repeated challenges that cause us to constantly examine and
re-adjust ourselves on our humanity—divinity tight rope. No one
becomes drunk by just reading the label on the wine bottle. We
must taste reality from moment to moment. That's what we're
here to do.

During the 1960's the "love generation" sought to utilize it's
idealism, concern for humanity, and desire for spiritual growth to
open to that divinity part of themselves which we have been talking about. However they attempted to do this by pushing away the
everyday world of forms and the learning experiences that it offered. That world was a downer, so by getting high they could transcend it. They would, through natural experience or the use of
chemicals, be transported into the divine realms of oneness and
spacious awareness and then ultimately "come down" back to the
everyday world of form. They would then escape from the forms by
again getting high and then they would ultimately come down.
This roller coaster process provided a glimpse of the greater part of
awareness beyond the forms but was transitory and unfulfilling
because it didn't last. By not lasting it acquired one of the characteristics of the world of form, namely being constantly in change.
It was not eternal or indestructible; rather it was just another form
of addiction that seduced the traveler into wanting more and more
of the same. It was another, subtler form of suffering because it
emanated from clinging of the mind to getting high and staying
there.

Opening up to the greater part of one's own beingness can be so seductive we can become addicted to it. This type of addiction is one aspect of habitual alcohol and drug abuse. Aside from the usual reasons of escaping from one's perceived problems, physiological cravings, and psychological dysfunctions — there are chemical openings that present views of our divinity, from which we don't want to go back to mundane earthiness and constant growth lessons of our everyday lives. We become trapped in not wanting the optimum sensations to lessen, and we attempt to stay "up there" through excessive use of the substances that apparently took us to these places beyond daily waking consciousness.

So the idea, is not to get high, but to *get free*—with no mental attachments at all to the nature or quality of our experience. One doesn't get free by becoming lost in any aspect of form, whether they be the usually serious dramas of manifestation, the 60's dramas of spiritual tripping, or the escapism of alcohol or drug abuse. Getting free requires an understanding that the only thing that brings one down is the clinging nature of one's own mind to staying high. And in these mental attachments which are the downers, are found the clues to one's emotional buttons that reveal where the mind is clinging and where our real work of attitudinal existence lies. This is really a gift from the Universe that always shows where work remains to be done. So the task at hand is not to push away the experiences or withdraw from the world of form, but rather to welcome them and consume them as the food of freedom. *Freedom is experiencing the fullness of life* in each moment to moment, including the suffering and the joy, the agony and the ecstasy, the aspects of the polar extremes and everything in between.

THE CIRCULAR JOURNEY

All the spiritual practices are primarily designed to free us from the clinging of our minds and the attachment to our own separateness, so that we can experience the underlying oneness of the greater part of ourselves. This is the oneness that births the forms of duality and separateness, to which we seek to return through the curriculum of our everyday experiences in the world of form.

This is a circular journey where we are constantly dancing from our unique separateness (the duality of the two) back into the totality of the one; then from the oneness back into our separateness and vice-versa. This is happening continuously and simultaneously. We go from the one down to the two and then back to the one. In the one that lies behind it all we dance as the two as we honor our incarnation of unique separateness. As we experience the world of form duality it continuously feeds us back into the greater awareness (oneness) that lies behind it.

As we start to complete the circle, the stuff of the world that used to be real for us now becomes just another aspect of reality. We understand that it is our curriculum for learning to go beyond it. This drama of the world of forms takes us out and back up to the one, and then the experience of oneness with its expanded awareness makes us appreciate the exquisite beauty and purpose of the world of forms so we come back down to the world of two for more learning experiences. This is the fluid process of waltzing through life—the dance of consciousness that we have been talking about.

We go down into the water and hold our breath in the world of forms and then we come up for air into the oneness space of our greater awareness. We go to sleep in the world of forms and then we awaken out of them. We are rising and falling, going in and out, up and down, back and forth constantly. As we appreciate and understand this circular process the purpose and meaning of our life changes and becomes much clearer. I realize that if I play skillfully enough in the world of forms and use them as my curriculum to free myself from attachment to them, I can truly delight in this incarnation. In using my dramas as the vehicle to free myself, my world is both incredibly full of sensations and totally empty of attachment to them at the same time and my experience of life is a delightful dance.

We begin to acquire a sense of the perfection of it all when we stand back from the dramas and notice them as a detached observer. We see that the phenomenons of life such as sickness, old age, and even death are mind-body processes which are conceptual mind labels as either good or bad, wonderful or horrible, positive or negative. From this vantage point we see everything being related to everything else in the game of life. It includes our layering, unlayering and relayering processes on the multilevels of existence and the honoring of our divinity and humanity. We are fluid vessels, constantly filling and emptying with the stuff of life. We are constantly going in and out, up and down, remembering and forgetting, clinging and letting go — over and over and over again. It's a process of gradual awakening of our awareness where we waltz our way through one day at a time.

As we do this dance and continue to expand the spaciousness of our awareness we meet other beings who are also dancing to the same music and not being lost in their roles. All of us are learning to offer the spaciousness of our own ever expanding awareness to others. That in turn allows them to be whoever they need to be wherever they are on their Journey. We are, after all, all seekers of our own truths. This creates further openings of awareness and the entire process is one where as we change ourself for the better we are changing our world for the better. As we view the perfection of the game we have a strong sense of trust that we are connected to and dancing in a loving and benevolent Universe. This Universe bends in our direction. It is for us and not against us. There are not errors or accidents in it. It gives us exactly what we want and need in order to sustain ourselves in our Journey of growth and awakening. It is as loving and compassionate, kind and caring, gentle and humorous as it is teaching us to become.

So we see that the purpose of the game is not to deny the world of form and push it away, either by rigid spiritual disciplines or through avoiding taking the curriculum of everyday experience. Neither is it appropriate to prematurely ascend out of the world of form before the necessary lessons have been learned. Our amusement park for this visit to the earth plane is right here and now each moment to moment, including the roller coaster rides, the scary side shows, the indigestion from eating too much cotton candy, and the laughter of the clowns. We're talking about the physical circus and the spiritual circus all existing and interrelating at the same time and place. It is the real "Greatest Show on Earth." In other words, what we're supposed to be doing here is "What we're doing here."

PART III
HOW DOES THE UNIVERSE WORK?

Where is there any book of the law so clear
to each man as that written in his heart?
    —LEO TOLSTOY (The Chinese Pilot)

That which God writes on thy forehead, thou
wilt come to it.
               —THE KORAN

The moving finger writes; and having writ
Moves on; nor all your Piety nor Wit
Shall lure it back to cancel half a line.
        —EDWARD FITZGERALD
          (Rubaiyat of Omar Khayyam)

PART III
HOW DOES THE UNIVERSE WORK?

THE NATURAL LAWS

Twenty five years ago I was sworn in as an attorney—member of the Florida Bar. The idea of practicing law without knowing the governing legal principles would, of course, be ludicrous. However, this is exactly what we do during our visit to the earth plane. Our Journey may have rules of conduct relating to daily societal involvements, but we usually must fend for ourselves blindly using the painful trial and error approach when it comes to our personal quest for a life of harmony, contentment, peace of mind and happiness. However there are also rules for this part of the game of life, and they are all around us if we can gain the necessary awareness to tune into them. They permeate our existence, touching every aspect of our lives. As understanding of them can help us to understand our past, predict our future, and live our present with a greater sense of comfort, acceptance, and a childlike sense of awe and honor.

I have attempted to compile a comprehensive list of these "hidden" Universal Laws that are basic and essential to our attempts to live "the good life." They are the regulatory vibrations that maintain order in our lives—when we live within them life is positive, and when we transgress them we suffer.

How skillfully we achieve the delicate balance between the multilevels of our worlds of humanity and divinity depends in large part on whether we are acting in harmony with these Laws

41

that govern and control our Journey. They are not the rational laws of the intellect, or man-made judicial or scientific laws, as prescribed by a particular authority. Rather, they are the NATU-RAL LAWS that run the Universe, knowable by close attention to certain uniform patterns of occurrence observed in natural phenomena and the heightening of our own awareness. Our understanding of these Universal Laws and compliance with them determines whether our experience of life is happiness and joy, or pain and suffering. Indeed, as Plato said, "To know God is to know the Law."

These laws are part of the foundation of knowledge and information that is our intellectual bridge. We can then cross it by affirming physical experience to maximize our personal growth, enjoyment of life, and spiritual evolution. The laws are there behind it all. Trying to navigate our journey without an awareness of them is like trying to steer a boat without a rudder. We can no more skillfully accomplish our life's purposes without a knowledge, understanding and appreciation of the Universal laws as we can while we're still tightly layered, lost in identification with our personal dramas, and erroneously focusing our lenses of perception "out there" in the physical world of forms. As we keep peeling away the layers of our attachments which have restricted our movement and shielded our vision, we get glimpses not only of our connections to the greater part of our conscious awareness which is one of our overall goals, but also clues to the system of law in which we have been blindly stumbling around. When things are going well in our visit to this earth plane we're not too concerned with the rules of the road. But when we lose our way or have a head on collision with the clingings of our mind—they are much more important to us. If we are attentive, the rest of our trip can be much more enjoyable.

By focusing our attention and heightening our awareness we see that everything is perfectly lawful in our lives and the world in which we live. On its face this seems like quite a mouthful, an outrageous statement to say the least. How can things be "perfect" with all the emotional roller coaster rides, soap opera and suffering that seems to pervade human existence? The perfection is seen when we are able to take a deep breath and move one step back from our dramas in the world of form and gain the elevated perspective from the greater part of ourselves.

The Universal laws can help us to understand this perfection of karmic and personal interrelationships that exists all around us. They are, although not observable or measurable, every bit as valid as the laws of science. Everything in our world is lawfully interrelated, with the Universal laws being the corner stone of all knowledge.

In fact, we become aware of these natural laws every time we violate them. Unlike man-made or scientific laws, where one could be totally unaware of how they work and still live reasonably happy lives—transgression of Universal law directly impacts us *every time*. The proofs are in our feelings of being out of synch, having what we call bad luck, disease, continued experiences of negativity and suffering, heavy dramas, and multiple other aspects of "going against the grain" of life. If we could know these laws in advance, think how much more skillfully we could navigate; and how much more enjoyable our journey would be!

1. The Law of Forms

One of the many paradoxes in our Journey of Awakening is the fact that "the most constant thing in life is change, and our resistance to it." The concepts of solidity, everlastingness, and indestructibility are not applicable to our three dimensional world of form. In fact they are just all thought form creations of our ego thinking mind which is also an aspect of form, inhabits the same realm and is subject to the same limitations. Everything is in actuality a perpetual movement of mind-body processes in constant ebb and flow change. This includes the pulsating energy that is the basis for our Universe and all its creations, as well as the protons, electrons and other subatomic particles that vibrate so rapidly they present to our physical senses the illusion of solidity and non-change. Our beliefs, thoughts, emotions and the actualized experiences which they create are all subject to the same law of change that constantly move us in and out of our experiences. Our morning of despair can become our afternoon of rapture, in the blink of these pulsations of swirling energy, because all the forms are just a parade of neutral phenomena. They are an interplay of light and shadow—patterns of energy focused in physical form—judged and evaluated by our ego thinking minds.

2. The Law of Perfection

We are all born as perfect beings. There is nothing we need to do in order to achieve physical, emotional, or spiritual perfection. It is our birthright from the Universe. Our perceptions of lack and limitation, fear and anxiety, pain and suffering, are learned illusions which can be discreated by appropriate methods in order to reveal our underlying perfection. This can be exemplified by the statement that, "We are already enlightened beings, pretending not to be." We labor under a condition of temporary amnesia which is always curable by expanding our moment to moment awareness to view the greater part of our beingness.

3. The Law of Energy

With the introduction of Einstein's theory of relativity, the age of Newtonian physics ended and it was scientifically accepted that energy created matter, rather than vice-versa. The entire Universe including ourselves, is moving energy that travels at various vibrational levels and creates, modifies,and discreates the world of forms. The multi-leveled aspect of the Universe is a reflection of the differing vibrational frequencies that have been achieved throughout it. Energy is perpetual and indestructible; therefore, it is not subject to the functions of the world of form such as birth and death; rather it continuously transforms itself to higher or lower vibrational frequencies. Energy never stands still. It either expands (creates) as Yang positive energy or dissipates (transforms) as yin negative energy. Since the flow of energy constantly changes, no adverse consequences are experienced unless one intervenes with it's free movement.

Imbalances in the flow of energy through the physical body vehicle or blockages which prevent it's free flow are manifested in emotional feelings of disharmony and disorientation, and physical symptoms of pain and illness. The term dis-ease is derived from these interruptions in the balance of our energy flow.

4. The Law of Patterns

This is a corollary to the Law of Energy which provides that the energy fields comprising and surrounding us contain all accumulated wisdom and experience. They have identifiable and patterned criteria which can be tapped into to reveal past, present and future significances through psychic analysis and divination. The analysis of particular energy patterns is categorized from a metaphysical standpoint into: (a) channeling mediumship (b) astrology, palmistry, and numerology (c) clairvoyancy and psychic healing (d) card reading, rune casting and I Ching practice. Modern psychology refers to this energy repository as "The collective unconscious," and modern science talks about it's commonality of shared experience as "the theory of morphogenetic fields."

5. The Law of Unity

The appearance of separateness in our physical world of form is another illusion perpetrated through our physical senses. In truth, "There is only one of us," interconnected and interrelated at the deepest levels of the Universe. We are all sparks of that energy essence; drops of water in the common ocean of conscious awareness. Because of this basic oneness that pervades the Universe, we each have a direct and proximate effect upon its vibrational rate. When we are disharmonious in thoughts, words, or deeds, we project this negativity out into the Universe and it decreases the overall vibrational rate. Since the object of existence is to move the Universal energy ever forward, raise its vibrational level and thus create additional energy, the heavier density of disharmony is counterproductive. Since every other being is also an integral part of Universal energy consciousness, our moment to moment actions and their underlying motivations have a similar proximate effect upon every other being.

6. The Law of Actualization

We are all the ultimate Creators of our own experience and our reciprocally shared realities. Everything we experience at the physical level has been manifested sequentially through our beliefs, thoughts, feelings and emotions. Our physical reality is a reflection of what we believe it to be. Our world is a reflection of our consciousness; and we can actualize any desired reality into our own physical experience through the unlimited creative power of our own minds. Our perception of people, places and things viewed through the filters of our belief systems determines our moment to moment experience of reality. Since we have the power and the ability to change our beliefs, thoughts, feelings and emotions at any point of our moment to moment existence—our experience of actualized physical reality is also similarly changeable. If we have a desire, believe in our power to physically bring it into being, and act accordingly with focused intensity—we will create the desired physical experience of reality.

7. The Law of Expectation

This law further relates to the power of our mental belief systems. Our physical world is truly a confirmation of the beliefs, desires and expectations that we have actualized. Since an expectation involves a desire for something to occur in the future and is a direct reflection of an underlying belief, we will always get what we expect to get. Things and events become for us exactly what we expect them to be. However, this may not be what we want or need if our expectations are not consistent with our underlying attitudinal beliefs. Any conflict interrupts the linkage of interconnectedness that enables us to create a desired reality. Clarity on all levels is essential in this process.

8. The Law of Mirrors

Just as our experienced physical reality mirrors our underlying beliefs, ideas and desires; so do we mirror each other in our interpersonal relationships. What we dislike in others, provides us clues as to parts of ourselves with which we are unhappy. Flaws in others usually mirror flaws in ourselves. This mirror effect provides us with a clue to suggesting that we turn our focus of awareness *outside-in* to reveal our own necessary areas of attitudinal change and healing.

9. The Law of Attraction

Similar to the mirror effect, we can only attract to ourselves qualities which we already possess. If we desire peace of mind, contentment, and love in our lives these attributes can only be drawn into us if we concentrate upon them and are in fact already calm, contented, and loving in our underlying structure of beliefs. The flow of desired qualities can only be energized through a common stream that attracts them back to its source.

10. The Law of Resistance

Just as we can attract that which we concentrate upon, we also draw to us that which we resist. We do so because our resistance is a manifestation of fear which is treated by the Universe as something that we will encounter continuously until we are forced to squarely face it and thus learn the necessary lessons which it represents. A person that we resist is drawn to us and we may in fact become that person in a future incarnation. An event that we resist will also be perpetuated in the same manner until it is consciously dealt with. Since all the lessons to be learned are attitudinal, once we gain the necessary awareness of the situation and we change our limiting perspectives that have heretofore blocked our necessary learnings, the "problem" dissolves, its energy is released back into the universal flow, and the experience becomes just another of our successfully completed opportunities for growth.

11. The Law of Opposites

All the Universe and its functions exist in a balance between polar opposites. As a part of our inherent unity, we contain dual aspects of humanity and divinity, Yin-negative and Yang-positive energy, beliefs in good and bad, feelings of love and hate, experiences of harmony and chaos. We live our lives in the creative tension between the opposite polarities, and it's energy is essential for the continued existence of our physical structure of form. The continuing interaction between the opposites generates the perpetual energy that is our spark of life. Since the polarities are essential for the maintenance of our existence, and since all of the forms which are created within the spectrum of our force field of consciousness are constantly changing, it is ridiculous to judge anything to be better or worse than anything else. Our viewpoint of highest consciousness must be moment to moment choiceless awareness; not this *or* that, but rather an acceptance of this *and* that as constituting the totality of life.

12. The Law of Paradox

Our physical world is not what it appears to be; it is a mass of contradictions that require one to go beyond the physical senses of the exterior world of form. Things may be relatively true, relatively real, and we may be relatively happy. There are no absolutes. Rather than our experiences being black or white, they are varying shades of grey depending upon the manner and method in which we react or respond to them. Examples of paradox include the following:

a) The only thing constant in life is change.
b) The goal of our Journey is the journey itself, rather than it's destination.
c) In order to experience happiness one must be happy; or the corollary that you learn to accept life as it comes by accepting life as it comes.

Numerous other examples are constantly being presented to us; this law requires us to look beyond appearances and to open ourselves to considering the greater possibilities of existence.

13. The Law of Visualization/Imagination
What we picture for ourselves as being true in the current moment is true as far as our subconscious mind is concerned. We physically experience the imagined reality in the same physiological manner as if it actually occurred, it is only our thinking mind which suggests to us that this is fantasy or imagination. Utilizing our powers of visualization through our world of imagination we can re-experience and rescript past events, create the present, and test out alternative possible future experiences.

14. The Law of Imitation
We can utilize our sensory powers of observation in order to model our behavior upon what has been observed. Positive observations inspire imitation and the opposite would also be true for negatives. We can utilize this law for the purpose of analyzing how our methods of dealing with and reacting to experiences may be a patterning of prior observation. When this awareness is gained, we can more skillfully create linkages of positive beliefs — emotions—experiences, and discreate unsatisfactory realities.

15. The Law of Benevolence
The Universe bends in our direction, it is for us and never against us. It wishes us to learn our lessons of incarnation in as

gentle and joyful a manner as possible. This law suggests that the best will find us if we allow it to. We have within us everything required to live a life of happiness if we choose to be aware of and accept the Universal abundance that is our birthright. A corollary is the law of opportunity which provides that when one door closes to us another automatically opens. When we take one step toward the Universe, it takes one hundred steps towards us.

We are constantly drawn toward our maximum growth. We are given repeated opportunities by the Universe to complete our lessons. As we finish one, another emerges, and the growth process continues. We don't have to do anything, just be aware of it, and accept it as it unfolds. The process is gentle, unless we resist.

16. The Law of Synchronicity

The Universe constantly provides us with meaningful coincidences to indicate whether or not we are in harmony with its flow of energy. These clues are a part of the constant communication that is offered us by the Universe to assist us in maximizing our experiences of learning, growth, and happiness; as well as being constantly reassured of it's benevolence towards us. Examples are:

a) I drive to the expressway toll booth, have no money and notice the necessary change on the floormat of my car.

b) I want to phone or write a friend but I have lost the number or address. At that moment they call or a letter from them arrives.

c) I need to contact a reasonable and fast house painter. In the driving lane next to me I see a panel truck that advertises "Paint your house—Cheap, one-day service" with the phone number for contact.

17. The Law of Serendipity

The Universe further evidences the extent to which it bends in our direction by providing us with seemingly accidental and unexpected fortuitous circumstances which we call good luck. These clues to the cosmic largess can include;

a) I entered a sweepstakes contest many months ago, forgot about it entirely, and on the day my income taxes are due I receive a check representing my winnings.

b) I've forgotten where I parked my car in my garage and the elevator opens to the correct floor while I'm still trying to remember it.
c) I'm at the race track, I inadvertently purchase a ticket on the wrong horse, and it wins.
d) I'm late for an appointment at an office building and as I circle the busy block a car vacates the perfect space for me right next to the front door.

To a certain extent all sychronicity experiences are serendipitous because they all carry the aspect of providential good fortune.

18. The Law of Experiential Perfection

Every experience we have on the physical plane is perfect from the standpoint of our growth process. When we are comfortable, at ease, contented, and happy (when we are in a positive state) we are in harmony with the Universe and our life flows accordingly. When, on the other hand, we are agitated, upset, anxious, depressed or disoriented (we are experiencing negative states of existence)—we are receiving the clues necessary to remind us that we are out of balance and attitudinal adjustment is necessary to center us once again. This is a perfect self-regulating mechanism which allows us to continually fine tune ourselves, as long as we have the awareness to notice it and act accordingly.

19. The Law of Altruism

When we help and serve others we raise our conscious awareness and our vibrational frequency to its highest levels. This is an automatic centering device where if we are already in harmony when we perform an act of selflessness we will exponentially increase those feelings, and if we are disharmonious at the time of the selfless act we automatically will balance any blocked energies and will regain our state of harmony and raise our energy level as a result of the act. The more we give, the greater is the increased flow of our energy and the more we receive back. The more we participate in acts of altruism, the greater are the multiples of increased positive energy that result and the more we experience the beneficial consequences of such actions.

20. The Law of Quietness

The most important communications between the greater part of ourselves and our physical sensory mechanism are the quietest. Our high intuitive self, the entity that communicates with the greater part of our conscious awareness, speaks through the still, small voice within. The more we quiet ourselves by reducing our mental chatter, increasing our awareness of our multi-leveled/multi-layered existence, and taking time to notice what is going on around us (stop, look and listen)—the louder the quiet inner dialogue becomes. Our impulses toward quieting down come from the higher self. They are sent to us in an effort to reduce the disharmonious and distracting levels of noise that occur from our awareness being focused on our exterior material world.

21. The Law of Repetition

The Universe continually presents us with the experiences needed to maximize our opportunities for growth. These lessons have been predetermined as our karmic blueprint. The learning of them is the purpose of our physical incarnation, utilizing our everyday experiences in the physical world of form as the curriculum for learning. To the extent we do not pass our test of experience and fail to achieve the degree of learning awareness necessary, the physical experiences which present the lessons are revisited upon us. The drama scenarios are repeated over and over again until the satisfactory learning occurs. Whether we avoid the learning experience entirely or deal with it imperfectly, the result is a replay of the basic blueprint. Once we pass the lesson, it is not repeated in this lifetime nor in any subsequent lifetime during which we take physical incarnation because we work on other lessons of soul evolution.

22. The Law of Karma and Reincarnation

This is the most fundamental law governing the nature and purpose of our existence. It is the rebound theory of physics—the energy we send out comes back to us like a boomerang.

It is also described as the law of cause and effect, self-responsibility, and Universal harmony. All of our words, actions and deeds and our motives, intents and desires behind them have a direct cause and effect relationship. Our disharmonious acts are like throwing a rock into a still forest pool and disturbing its harmony. The ripples that result from the splash in the pool flow out and back until harmony is eventually restored. Similarly, our negative karmic actions flow out into the Universe and reverberate through numerous lifetimes until eventually the required harmony is restored by the effects of the karmic act being balanced. We reap that which we sow, and therefore all that we bring into the lives of others comes back into our own. Our physical lifetimes represent efforts to learn the necessary lessons to even up the balance sheets of past karmic acts so that ultimately when we have cancelled out all the negative—disharmonious karmic experiences by having balanced them in a positive and harmonious experiential way, the harmony of our personal cosmic pond will have been restored and it will not be necessary for us to physically incarnate any longer. The key is getting free from the cycle of physical births by achieving total harmony in all areas.

We have free will and choice as to the specific manner in which we experience the required learning of the karmic lessons; either through joy or through suffering. The higher our awareness in this lifetime and the more consciously we live our lives will determine the extent to which we can lessen or erase the karmic debts that we brought with us into this physical incarnation. It is possible to balance our karmic debts in a particular lifetime, without having to fully experience the blueprinted intensity of the dramas that were pre-planned on the soul level prior to incarnation—by a combination of the erasing of negative karmic balances and the accruing of positive karma through a lifetime of higher consciousness, helping and serving, and practicing unconditional love and acceptance.

23. The Law of Attitude

Since we create our own physical reality by actualizing our belief systems and the feelings-emotions which they generate, it is our outlook that determines whether or not we have a problem. At this attitudinal perspective, from one moment to the next the external facts and circumstances of the situation may remain the same but, whereas we may have viewed it as a problem and experienced suffering one moment, the modifying of our perspective about the situation in the next moment can result in it no longer having any affect upon us. This also confirms that our point of power over the circumstances and quality of our lives is always in the present moment. We are the creators of our life experience, the internal cause rather than the external effect.

24. The Law of Choice

We always have free will and choice in our response to any external situations. We are never powerless because there are always alternative courses of action which result from the internal attitudinal nature of our moment to moment experience of life. We can change our perceptions of people, places and things at any time. In addition, on the soul level, prior to our incarnation in the world of form, we chose the astrological configurations of birth which may strongly determine our personalities, physical and emotional characteristics, and potential areas of strength and weakness. We also chose our parents, the circumstances of our birth, and we created our soul blueprint of the major learning experiences to be encountered during this visit to the earth plane. Through our abilities to materialize our own reality, we also choose and create all physical events which we encounter in our lives. Therefore, there are no accidents, no errors in the game. Everything is as it should be because it is all our creation in the first place.

25. The Law of Intensity

The more focused our directedness toward a desired reality, the more likely it is to be actualized. The clearer our awareness and understanding of what we view as our goal, and the decisiveness in which we set out to accomplish it creates a tremendous whirlpool of creative energy that seeks to form the desired manifestation. This intensity of directed emotional energy is like a laser beam. It acts as a magnet which attracts to us that which we seek. It is an integral part of the process of reality creation.

26. The Law of Acceptance

All suffering and experience which we conceptually label as negative is the result of our failure to accept what is. We can utilize our best efforts to create desired realities and discreate undesirable ones. However, once the situation is clear that the desired creation/discreation is not in the way of things, we are furnished another clue from the Universe that we must emotionally detach ourselves from the intense energy which we had previously invested in the situation, accept it, and move on. This is a law of happiness in our daily experience of day to day living. We can change what is changeable, but we must accept what is not, and redirect the energy into other more appropriate areas. This allows us to enjoy all positive aspects of our lives, but to be able to let go and create an opening of detachment through which any negativity will flow through us without harmful effect. This negative energy can be utilized through new creative challenges in our lives, and be converted into positive energy.

27. The Law of Simplicity

Simply stated this law means that "less is more." Our most important learnings in life are profound in their simplicity. Rather than looking for complicated answers and involved procedures to solve our problems and modify an apparent negative life experience, one needs to focus the lens of attitudinal perception from the outside where the noise and clutter and drama of life is occurring, back to the inside where one's still, small voice of the all knowing higher self can provide wisdom, peace of mind, and true happiness. Another Universal clue occurs when our lives become so busy and filled with external activity, we physically and emotionally experience symptoms which strongly suggest to us that it is time to simplify by stopping, quieting down, and refocusing inward. The greater the exteriority of our focus, the more complicated and cluttered will be our lives. Our experience of life will be disharmonious, and we will have strayed from our optimum path of awakening. When we skillfully respond to this situation with awareness and corrective action, we receive confirming emotional and physical clues that we are back on the right path, continuing our journey.

28. The Law of Here and Now

Our lives are a perpetual series of "now" moments. The tendency to cling to the past and grasp for the future is just an erroneous preoccupation with the idea of linear time. In actuality life is lived to the fullest in each moment to moment of choiceless awareness. Letting go of the past and trusting in the future are keys to allow a more direct current focus, and provide the experience of happiness.

PART IV
WHY DO WE SUFFER?

He has seen but half the Universe who never
has been shewn the house of pain.
>   —*RALPH WALDO EMERSON*

I count him braver who overcomes his desires
than him who conquers his enemies; for the
hardest victory is the victory over self.
>   —*ARISTOTLE (Stobaeus)*

No conflict is so severe as his who labors
to subdue himself.
>   —*THOMAS A. KEMPIS*
>   *(Imitation of Christ)*

PART IV
WHY DO WE SUFFER?

"If I have a desire, on whose fulfillment my happiness depends—it is in reality a chain around my neck and I will most definitely suffer."

Anthony de Mello

Probably the most pervasive area of our shared common experience as human beings is that of suffering. We have all experienced it and we can all identify with it. In fact our very way of life, our personal dramas, our media soap operas, the blaring newspaper headlines and TV bulletins, and the most vividly remembered experiences of our lives are generally in terms of the pain or distress involved.

A current rock and roll song says, "I want a girl who won't drive me crazy." How often in the course of our daily conversations and interactions with others we say or hear, "You make me so mad, look what you've done to me, it's all your fault, you've ruined my day," ad infinitum, ad nauseam. But who is really doing it to whom? The answer is obvious; we do it to ourselves and we always have been. We incarcerate ourselves in prisons of our own thinking minds, surrounded by the nets of our sticky thought forms which seem to continuously entrap us. In the words of Bo Lozoff, "We're all doing hard time until we find freedom within ourselves." This suggests that we have internal power and choice over the nature and extent of our suffering.

THE CAUSE OF SUFFERING

Initially, all suffering is a direct result of our violations of one or more of the Universal laws. Thus, the better our understanding of them, the less the likelihood of our negatively impacting them. However, in those instances where we do transgress, either consciously or unconsciously, the intensity of our suffering and its duration is self-determined.

If the Buddha had been born in contemporary times, he would probably be on the personal growth workshop circuit opening his lectures as follows:

> "Good evening ladies and gentlemen. I've got good news and bad news. Which would you like to hear first? *Give us the bad news.* I thought you'd say that, your negative conditioning being what it is. Well the bad news and the good news are both a part of my four noble truths, which tell you everything you need to know about living a happy life. Wow, this room certainly got very quiet—I obviously have your undivided attention."

Here they are restated in contemporary terms:

1) the nature of life as a human being on the earth plane of form contains suffering.
2) the causes of suffering are the clinging desires of one's own thinking mind, holding on to models of how things should or ought to be.
3) there is a way to cease these attachments of mind that cause suffering.
4) the ending of suffering is attained by the Noble Eightfold Path which is a way of conscious, ethical living that will purify and harmonize body, mind, and spirit.

There are similar insights in the personal growth and human potential movement, gathered from eastern philosophy, western psychology, and updated into a contemporary context—all of which leads back to the wisdom of the ancients. The Buddhist Dhammapada says, "All that we are is the result of what we have thought; our existence is founded on our thoughts." The states of mind that create the five hindrances to happiness are sense desires, anger, laziness, restlessness and doubt. The Hindu Bhagavad Gita tells us to act but not identify with ourselves as the actor, or our actions when it says, "Do what you do, but be not attached to the results." Other ancient scriptures make the same statements. Our suffering relates to the emotional attachments of our mind and it's perceiving of our experience as good or bad.

Modern authorities say the same thing, just using different terminology. They emphasize that all suffering is attitudinal. How we react or respond to a particular situation is the key. "Reacting" implies a knee-jerk immediacy where we dive right into the drama, reserving no time for calm reflection and prior evaluation. We try to manipulate the Universe to give us what we want. "Responding" is the opposite; we have the awareness of what appears to be happening, but it doesn't blow us away. Rather, we deal with it in an appropriate and more skillful way after we understand it's dynamics. We try to manipulate our mind to accept whatever comes our way.

Thad Golas talks about mental resistance to what is and it's corollary which is letting go and acceptance. Ken Keyes talks about the only cause of our suffering being "addictions," which he defines as the emotion backed demands of our thinking minds. The positive alternative is to become aware of our addictions, identify them, and then upgrade them into non-painful preferences by utilizing various methods of mental reprogramming. Ken Wilber talks about our mental patterning which creates arbitrary separating boundaries in our experience. His "no boundaries" approach opens one to the greater part of self which lies beyond the thinking mind and is therefore outside suffering which is a product of the mind. Ram Dass talks about the "sticky mind" and how we constantly become lost in our personal dramas and suffer until we are able to extricate ourselves and "come up for air." He readily admits that he still has all his neuroses; now, however, he perceives them and their interactions with his world of form in much more skillful and responsive ways. Jack Schwartz similarly talks about how we carry our traumas, even the pre-natal ones, with us our whole lives. It is our "interpretation" of them in our daily experience that determines the quality of our lives. Harry Palmer, in his creativism concepts, emphasizes how our world is a mirror of our consciousness. He says, "Reality is a reflection of what we believe is real," and our beliefs create our experience. Our beliefs are the product of our thinking mind. Just as we have created them and habitually carried them and their resultant actualized life experiences with us, we can discreate an undesirable reality by changing our beliefs and recreate a more positive one. Patricia Sun says to own your feelings and underlying beliefs by fully experiencing them. Then and only then can we step back from them and be aware of how they are running us.

These authorities also suggest we must learn to not only hold on tightly, but to let go lightly. The monkey story is illustrative of our predicament if we don't:

Long ago in India monkeys were a source of food because their population had increased. The villagers sought ways to capture them. They set traps which were baited with sweet nuts and fruit, the monkey's favorite food. The baited traps would be laid out at night and the villagers could hear the chattering of the monkeys grow to a loud noise as they approached, viewed the traps and then removed the bait; all without being caught themselves. This went on for some time, to the growing agitation of the villagers. Finally, in desperation, they decided to consult their wise elders who suggested that the villagers hollow out a coconut which would be chained to a nearby tree or other solid object. Inside the hollow would be placed the bait for the monkeys. The size of the hollow would be that of closed, clenched fist which was larger than if the hand was opened and the fingers extended. That night the traps were set and in the morning hundreds of monkeys had been captured, all of them having reached in to the coconut and grabbed the delicious morsels of nuts or fruit in their clenched fists. They continued to cling to it for fear that it would be lost. As long as their hand was closed into a fist they would be trapped, once they opened it and let go of the bait, they would be free.

There are numerous other examples; the important factor is that while our suffering is self-created we can also, with awareness and application of appropriate skillful methods (detailed in Part V), let go of it. Recurring key words or phrases such as clinging, attachments, resistance, addictions, limiting boundaries, undesirable beliefs, habitual patterns of negative thought and action —all suggest the self-directed nature of our suffering and point us in the direction of resolving it. The key in all instances is our very own thinking mind, and it contains all the bad news as well as all the good news. Our mind truly creates our world!

There are many levels of "mind," all layered together and contributing their share to our moment to moment experience of calm or chaos. Some that act out various parts in our mind play are:

A) *The Chatterer* is a type of monkey mind that noisily distracts us with comments on everything, like a gossipy neighbor who talks incessantly, drowning out our quiet inner voice of higher wisdom.

B) *The Judge* passes sentence as good or bad on everybody, everything and every experience. The severest punishment is usually dealt to ourselves. He is also called *The critic.*

C) *The Worrier* sees potential danger around every corner. We are told to trust no one and be constantly on guard for the direst of consequences.

D) *The Cynic* views life through dark, cloudy glasses; seeing selfishness as man's sole motivating factor. He is not quite as active as *The Egotist,* who views any threat to his existence as an act of war, and defends himself accordingly.

E) *The Brain* lives in the world of the rational intellect. He is cold, aloof and insensitive, viewing life with bored detachment.

F) *The Dramatist* lives solely in the world of emotions and melodrama. In this role, there is constant soap opera traversing the ups and downs of the daily roller coaster of life.

G) *The Sage* sits back atop his lofty viewpoint of experience, judgment and wisdom; accepting and enjoying everything, without conditions or attachment, as all the delightful flow of life. He sees it all, acknowledges it, doesn't get emotionally drawn in to it, and gently let's it all keep going.

WHEN AM I SUFFERING?

One of the many perfect clues to the mystery of our human existence is how easy it is to know that some event has "pushed our buttons" and we are experiencing what is commonly described as suffering. It is a manifestation of our physical form so that it is always translated to us through our physical senses. Any of the sensations that we commonly regard as negative are in actuality the experiencing of some degree of suffering. We can analyze and evaluate these experiences by seeing in which of the three following categories they fall:

A) I want something, but I don't get it. (grasping, grabbing, reaching, desiring, expecting)
B) I don't want something, but I do get it. (resisting, avoiding, dreading)
C) I want something and I get it, but it ultimately changes and doesn't last. (clinging, holding on, attachment)

These A, B, C scenarios are of critical importance since they cover the entire range of the human experience of suffering. Their product includes sensory clues such as rapid heart beat, dryness of mouth, elevation of pulse, a general bodily tightening and rigidity, disorientation, lack of mental clarity, localized body pain, headache, muscle tightness, queasy stomach, bowel syndrome, skin-rash, tremors of the body and extremities, eye tics, etc. If these types of sensory clues are disregarded or one's awareness is so low that they are not even noticed, they will ultimately degenerate into specific health problems of disease or bodily disorder in order to direct one's attention to the necessary areas of personal learning that they symbolize.

There are a similar list of emotional symptoms of suffering that our mind labels for us when the physical sensory phenomenon occurs. Remember that our thinking mind is conceptual. We see or experience a phenomena which is merely a neutral flow of mind-body processes. Our conceptual mind simultaneously reaches out and gives it meaning. For example we hold up our foot and our physical senses see a combination of light and shadow and shape and our conceptual mind immediately gives it a meaning and creates it as a "foot." In the same way, in response to one of the A, B, C scenarios our physical senses produce a pounding heart, throbbing head or churning stomach and our conceptual mind immediately translates this into an emotion such as fear, anxiety, jealousy, frustration, anger, guilt, sadness, insecurity, revenge, resentment, blame, and all the other numerous "negative" emotions. Our brain is just another part of our physical sensory mechanism—it is our bio-computer which obeys the instructions that are given to it by the thinking mind and translates these commands into the appropriate bodily function. It is the thinking mind that creates our perception of the physical universe in which we function and through it's conceptualizing we view our experience of life as either positive or negative.

WHERE DOES THE SUFFERING COME FROM?

A) The Internal Battle

We have seen from Part I that the greater part of our being, called I, is the pure universal energy essence conscious awareness from which our little *i* is formed. This *i* is the physical form that is inhabited by our conscious awareness as it passes through this lifetime in its visit to the earth plane. Since the thinking mind is just another of our physical senses, although the subtlest of them, it's product which is thought is also grounded in form and that is where we run off the tracks. The very nature of our thinking mind is that of judging and labeling and evaluating the phenomena of our life experiences in relation to it's role as guardian of the exclusivity of its ego self. Just as the big I is the unitive realm of "there is only one of us" the little *i* is the domain of duality and separateness. The big I speaks in terms of neither this nor that, the key word is *and*. It is totally inclusive of all people places and things as being valuable opportunities for learning. The little *i* speaks of *either-or* and it's very nature is dualistic; judging everything in terms of whether or not there is a threat to its egoic existence.

The thinking mind prefers the status quo. Even though it recognizes the fact that "the only thing constant in life is change," it views this as academic information only and not applicable to limit it's authority. We encounter this particular form of "suffering" whenever we try to determine what is the appropriate form of expression for us at a particular time on the physical plane. The answer of course from our all knowing I is that whatever we're doing right now is the appropriate thing to do. From the *i* level however our predicament is that all forms are changing from moment to moment. Our thinking mind wants to be certain. It has a model of who or what we are and it wants to maintain it. This is a rigid and inflexible model, and since the world in which it exists is changing in each now moment, the model becomes our prison and the fluid world of form in which it exists is a constant threat to it's survival.

There is a similar battle stations approach of the thinking mind when it perceives any change in its role as the five-star general of our world. In reality the thinking mind is our servant. However, it has assumed the larger role of master and done such a good job in convincing itself and us — it meets any attempt to diminish it's over-extended sphere of influence by declaring war. As the battle rages, we spin around on the wheels of our chattering mind, caught in a whirlpool of discordant thoughts that bounce us between calm and chaos and translate into an experience of suffering.

A balanced perspective in assessing the mind's functions is essential. It is certainly important to our everyday existence; but it is not the whole show. It takes us to a certain point in our journey, like a booster rocket that falls away once the mother ship has been transported out of earth's dense atmosphere. We then can rely more on aspects of us to which we have opened, beyond the mind, to propel us to our true destination as we experience life on a much deeper level.

B) Shutting Down

The layering process that we talked about in Part I is itself usually physically and psychologically painful. As the pure energy essence of our conscious awareness (the big I) is covered by negative belief systems, our physical reality affirms and self generates this negativity. We experience physical sensations and emotions that are consistent with this constricting, desensitizing and de-energizing process. The more completely we shut down the more intense will be our pain and suffering. As the layering becomes more severe, the intuitive part of us realizes the impending damage, like a nuclear melt-down, and our warning buttons are pushed on deep psychological levels so that, while we may not experience specific unpleasant physical sensations, there is still a free floating anxiety that suggests to us that something is wrong even though we don't know what it is. This is another aspect of the ABC's of the cause of our suffering which may be stated as, "I don't know what it is, but I do know that something is wrong."

C) Identification

We can see that to the extent we identify with our personal dramas as "reality" we experience emotional suffering. They can be viewed as a wheel of life. There is a hub at the center and an outer rim encircling it. If we attach ourselves to the external rim, we will either be at the top going down or at the bottom going up. We will be constantly roller-coastering back and forth between these polarities. However,if we stay on the hub, we will be centered at all times, no matter what direction the wheel turns.

Similarly, our identification with our physical form as "reality" produces suffering because it lives in the world of change; it is continuously in a state of decay and ultimately it is the part of us that dies. Our conscious awareness temporarily inhabits it (our space suit) during this lifetime, and releases from it when the body drops away. However, if we can live our soap operas without getting lost in them, we have the best of all physical worlds; the excitement and spice of life they furnish together with the overview awareness of our ability to detach from them. We can see how the whole game interrelates.

D) Polarities

The Universe is a great ocean of paradox. It's boundaries are defined by bi-polar opposites. The space in between is neutral phenomena.

Part of our suffering is caused by our thrashing movement from one pole to the other—our either/or reactivity to people, places and things. We ultimately realize that happiness is disengaging ourselves from the magnetic pull of these extremes of good and bad, pleasure and pain, high and low—and allowing ourselves to float in the in-betweens, letting the flow of life carry us where ever it does, and trusting in the rightness of our destination.

Like the drive train of an automobile with its interlocking gears, whenever we engage the extremes we have pushed the gearshift to forward or reverse and we experience the ups and downs of our personal roller coaster ride. When we are in neutral, the gears are disengaged and we have the opportunity to observe our predicament.

If we hold on anywhere we accumulate desires, attachments, grasping—resisting—clinging. Paradoxically though, in order to let go of anything we first have to be holding on to it. Therefore, a certain amount of initial accumulation of "stuff" is necessary so we have a basis of comparison between attachment and non-attachment. First, we engage life and in so doing create a working bank of reactions, desires, and attachments of the mind. These are our dramas that provide creative tension, and power us between the extremes of our defined life experience. Next, we disengage and rest in neutral for a while, so we can analyze and process the experience we have just had. Then we re-engage and do it all over again to have more experience in form, analyze the new levels of our emotional reactivity, and once again disengage, become detached and evaluate what has just occurred from the neutral perspective of an impartial observer of ourself as we continue the cycle.

In time though, the ability to engage and disengage is cultivated so that it becomes a volititional skill where we can remove our focus of conscious awareness when we choose and shine it elsewhere, transporting oneself from reactivity to responsiveness; not bouncing between the polar extremes, but *floating in the neutrality of the consciousness awareness space that they define as outer boundaries.*

Like the Japanese art of Aikido—if there is no resistance, there can be no conflict. So a key to emotional resolution is floating in the neutrality of the space of "neither this nor that," rather than the combative duality of either/or."

E) Past Lives

The law of karma, which is the principle of cause and effect, determines the type, intensity and repetitive nature of our learning experiences from lifetime to lifetime. Sometimes our opportunities for growth, our learning experiences, are more intense than would ordinarily be the case because the necessary balancing of karmic acts is required. The very intensity of these experi-

ences, which may be repetitive throughout a large part of this lifetime, is again a clue to the possibility that they are karmic and
must be allowed to run their course. Unusually strong, irrational
fears and extreme reactivity to particular situations are almost
sure indicators of karmic suffering. This fact of awareness may in
reality be the karmic lesson; and its realization may serve to alleviate the suffering because with one's awareness of the karmic actions comes acceptance and non-resistance to their consequences.

The results of our past lives and how well we understood the necessary learning experiences are the manifestations of bad karma
or good karma that are reflected in the major learning experiences
which are our karmic blueprint for this current lifetime. This law
of cause and effect is such that what we create in our current experience on the earth plane is the effect of a whole long set of the
previous circumstances that just keep running on until they are
resolved in one way or the other. If we do not have this awareness
and we become lost in our dramas which are the effects of prior
cause and effect situations, we just become the cause of the next
effect, and soon we continue to generate more negative karma. It
becomes a self generating process. This karmic category is a form
of encoded potential for suffering that we bring into this lifetime;
however, our level of awareness and the way in which we experience our karmic blueprint (either through resistance and attachment or through acceptance and grace) determines in large degree
whether or not this disharmonious potential is actualized.

F) Societal Aspects

The societal conditioning of our western world with it's Judeo-
Christian roots also generates a strong potential for suffering. It is
the primary factor in the acquiring of our negative belief systems
and our destructive layering process. In reality we are human animals who have a direct linkage to nature and the land. Our western world suggests that man must conquer nature, rather than
exist harmoniously and be interconnected with it. Our emphasis
on industry, the external sciences, goal orientation, and the rise of
technology as a part of our emphasis on doing and achieving have
further contributed to the denial of our roots of divinity. This
again causes free floating anxiety and the particular form of suf-

fering that has become identified with our contemporary urban life and our solitary, insulated and fearful existence — *stress*. In this country we are the product of the puritan work ethic, a belief system that has become an integral part of our society and our existence. It suggests that one must work hard, be productive, realize that life is difficult, and expect suffering to be a natural product of the life experience.

We are inundated by negativity and examples of suffering in our everyday exposure to media. With the widening of our access to world wide events and the breakneck pace of living in the urban market place, our world view on a daily basis is that of suffering; personal, our neighbors and the world. Even our forms of recreational escape from these pressures of everyday life are books, movies, and media entertainment that feature dramas and soap operas whose popularity is measured by the intensity of the personal suffering involved. We eagerly input these "real life stories" because we have identified completely with our own dramas and are used to getting lost in them.

We often believe that our life is boring unless we create and recreate similar personal movies whose main theme is suffering. Dramas emphasizing happiness, love, joy, kindness and caring, compassion, and happy endings are considered to be the realm of fairytales and childishly unreal. This societal negative conditioning is so complete that we even find ourselves, when things are going well and we are experiencing an infrequent interlude of calm peace of mind, waiting for the other shoe to drop — almost welcoming the recurrence of a negative experience that we view as inevitable. These repetitive personal dramas are viewed as an accepted part of our human beingness. Since we also have an anthropocentric view of our universe, where man is the central fact and final aim of creation, it naturally follows that our experience of it is often a self-centered, drama-filled valley of tears.

The following comparison listing of 25 major distinctions between eastern and western philosophical — theological world views further illustrates the nature and extent of our societal conditioning which prepares us to experience suffering:

Western Philosophy (Judeo-Christian Roots) Aristotle's Scientific Rationalism	**Eastern Philosophy (Hinduism-Buddhism Roots) The Path of Alternate Consciousness**
1. Issues of good and evil-obedience/disobedience to will of God	Issues of life and death—the Universe doesn't judge us, it just *is* the way of things
2. Separation between God and man	Primordial oneness of all creatures—interconnectedness
3. Time orientation—future "ought to be"	Time orientation—here and now "it's all O.K."
4. Value Oriented	Nature Oriented
5. The Holy is realized in the "ought to be"	The Holy is realized in something already here and now
6. Humanity/Divinity is Father-Son/Child	Humanity/Divinity is mother—Child/Son-Daughter
7. Route to God—Prayer to "Him"!	Route to God—Meditation within! (Quiet the mind)
8. Focus on External Material World of the Physical senses-demonstrable facts. The world is real.	Focus on multi-dimensional inner-world meta-physical. The world is illusion (Maya).
9. World defined by senses and logical rational mind	Unlimited view of reality, the invisible world, many levels of reality beyond senses
10. Dualistic—value system of opposites (either/or)	Monistic—neither this nor that
11. Left Brain—Rational; logical	Right Brain—intuitive, creative inspirational
12. Emphasis on external science and Technology	Journey inward—focus within

Western Philosophy	Eastern Philosophy
13. Goal/achievement orientation and logical planning for materialistic success. "Get the job done" Emphasis on *doing*, achieving!	Be in the world, but not of the world—inner fulfillment. Emphasis on *being*, inner harmony, serenity
14. Rational/consciousness—the ego	Ego-lessness (our ego keeps us separate)
15. Plan life to fulfill desires	Desire-lessness (our desires cause our suffering)
16. Logic, domain of common sense, we understand—then believe	World of unfamiliar requires *faith*, since it can't be explained by rational mind-we believe before we understand
17. Man must conquer nature	Man is connected and a part of nature
18. Anti-feeling or un-reason	Pro feeling, intuition
19. Clinging to life—fear of death. We are our physical bodies	Lack of fear—cycles of birth and death. Our soul essence "rents" the body form
20. One God—has a human personae, God is external to us	God dwells within us, as us! We're all part of "all that is"
21. God's Judgement—sins must be repented; emphasis on prayer, confession, punishment, one lifetime on Earth	Karma and reincarnation—multiple physical lives; Universe loves us; emphasis on meditation, inner purification
22. Complex, dependent on God's grace	Pragmatic, self-directing
23. Forcing—make things happen	Allowing—accept what is
24. Aggression may be justified	Non-violence (Ahimsa)
25. The old are viewed as no longer valuable or productive, and they are cast aside	Elders of society are honored as wise teachers

G) Experiential Aspects

By their very nature, experiences of pain and discomfort (which result from our ABC scenario) are more intensely remembered than those types of experience we call pleasant. They become burned in through deep layers of our consciousness because the energy necessary to gain our attention is greater than our average responsiveness to everyday occurrences. This is a part of our deep layering which the mind labels as negativity because it has disturbed the comfortable status quo which it cherishes. As the thinking mind has created our models of who and what we are and the nature of our expected experience, those models have been creating a corresponding physical reality for us without us being consciously aware of it. As our mind creates these models of our world, our required roles in it, and the roles of the other actors, our experiences confirm the mental pictures as we project the film on our screen of life.

As long as we remain entrapped within our thinking minds and identify with our dramas, we can't truly know our triggering belief systems or the reality that we are creating through them. It's only when we are capable of standing back from our dramas and moving a step away from our thinking minds that we can see the true picture and not be trapped within the thought forms that our mind is creating.

It is said that the thinker being proud of his mind is like the prisoner being proud of his cell. The more intensely that the learning experience must jar us, we see the deeper are it's layers—when we ultimately gain the awareness to start the awakening and unlayering process. As Scott Peck says, "You can't forget anything, you can just remember it without pain," and the deeper it is embedded the more stubborn it is in resisting being upleveled and discreated.

H) The Illusion of Separateness

One of the greatest contributors to the suffering model that we encounter is the perception of our limitations, helplessness, and mortality. These misperceptions directly relate to our Judeo-Christian upbringing as flawed beings who committed original sin, and have a God who is vengeful, judging and who resides somewhere outside of us. Whether or not we experience heaven or hell at the end of this short lifetime is totally dependent on God's grace which may, in his sole discretion, be granted us. Combining this with the fact that our world is defined through our physical

senses and our western logical rational minds, our emphasis on
the exterior world of material doing and the hostile and dangerous
environment in which we live—it's easy to understand our belief
systems and confirming emotions and experiences of loneliness,
isolation, disorientation, and fearfulness. All of these fall within
the broad category of our separateness. Its opposite is the unitive
experience of the greater part of our conscious awareness where
everything is a joyful flowing energy essence of oneness. When we
are feeling threatened, out of control, and our mind is churning
with negativity, it is very difficult to retain faith and trust that we
are connected to a loving greater aspect of our being.

1) Birth

At the soul level we experience separateness when we leave our
perfect state of amorphous oneness as the soul chooses to create
itself as a physical form on the earth plane. A split occurs in the
creation of a male or female incarnation so that we carry an encoded stance of incompleteness which is now fashionably described as looking for our soul mate. In this respect we are our own
soul mate whenever we complete ourself by balancing our male
and female energies together. Nevertheless, this initial division
may also be described as separateness.

Before physical birth occurs our first significant awareness of
physical existence is in the womb. During our time there we are
still connected to our spiritual energy essence, totally protected,
nourished and secure. We are still "home." Then after nine
months of gestation what was previously wonderfully safe becomes life-threatening in it's confinement and we are forcefully
thrust out of paradise. This is an intense experience of separation
which we carry throughout our lives at various levels of awareness.

At birth we have outgrown all that was our previous physical
Universe. We are expelled into a new world, heavy in the density of
our humanity, and all the related aspects of living in a physical
body here on earth. As we are propelled through the narrow birth
passage into the unknown new destination, we experience for the
first time feelings of constriction, struggle and fear. The pain of
this birth experience imprints deeply within us the desire to return "home" to the safety and blissful familiarity of the womb

state — our lifeline to the greater part of our spiritual energy essence.

After birth occurs we retain our linkage to "home" through the nurturing of our mother, but since the growth process is inexorable we must ultimately part from her, establish our own identify and begin to live a physically separate existence. This also generates suffering since we have a deep desire to grow, expand and move forward; yet at the same time we feel a deep sense of loss at what we leave behind. *We experience the joy of expansion and the pain of loss simultaneously.* At the same time we feel compelled to move along life's highway in our physical incarnation, we also long to return "home" and merge with our primordial oneness.

2) Sleep

We again experience this suffering of separateness, an almost genetic sense of having lost our greater half, in our sleep state when we return "home" to our unlimited Universe where time, space and the other limitations of our world of form are nonexistent. There we are replenished and revitalized, our problems can be resolved, and we receive meaningful insights into our daily dramas. We get to test alternate courses of action, visit with friends and loved ones, see glimpses of the future and in many ways revisit paradise. We can even take reminders with us back into our waking state through dream recall and understanding of dream symbols.

But when we wake up we have again the sense that we were thrust out of our "home" as we return to our physical bodies and our continuing personal dramas. We often feel the familiar anxieties, incompleteness and longing to return to what we just left. The peace and tranquility of this greater state of our being can be addictive in it's allure when compared to our daily life of churning minds, heavy emotions, and sticky problems.

3) Sensory

We also experience physical separateness that is constantly reinforced by our visual senses. Everywhere we look, we are all doing the dance of life in different physical forms. We look, speak and act differently. Duality seems to be everywhere and our senses verify it constantly and send us messages of affirmation. Since we live within the confines of our thinking minds, we accept these communications of separate phenomena and develop our mental concepts of "me and you, us and them, mine and yours, friend and foe." By repeating these speech patterns we further reinforce the illusions of separateness and isolation and act them out as affirming experiences. It all becomes a self-perpetuating cycle of separateness and suffering.

The nature of our life styles in the contemporary western world is also an exercise in separateness. The nuclear family usually lives alone in isolated box like dwellings, insulated from contact with one's neighbors. Gone are the days of the extended family, with its mutual support and nurturing. As we "do our own thing" we usually act alone rather than in cooperation with others. The pace and style of our lives in striving to get ahead and succeed at all costs to achieve those symbols of material success located "out there" mitigates against meaningful personal interactions and sharing of consciousness with other beings on an elevated level of awareness. The notion is that spirituality and material success don't mix and one cannot exist with the other. This either-or duality pulls us away from our inner desire to connect with the greater part of our being that exists in our lost higher self.

These are all examples of our separation from our essence. The big I and the little *i* are an integrative experience. To the extent we are out of balance with one or the other our disharmony triggers another ABC scenario of mental suffering. Again, our heightened awareness, acceptance of what is and letting go of rigid models of shoulds and oughts is in itself both the unifying and healing medicine, as well as the wellness state that is the desired cure.

THE PURPOSE OF MY SUFFERING

The suffering that we experience and create for ourselves is not a mandatory punishment that we are required to endure. Rather, it is a part of the perfectly lawful way in which the Universe has set up it's game plan. In addition to the required balancing of karmic debts, it provides for the maximization of our learning experiences and opportunities for personal growth. Our suffering gets our attention and requires us to focus on specific areas of learning that are highlighted by the painful experience. The suffering itself is a clue to the areas in which our minds are clinging, our mental attachments exist and where we need to focus our attention—they are like a flashlight beam illuminating hidden areas that we need to explore. They push us to make the necessary changes in our life that we otherwise resist due to our limiting belief systems and negative conditioning. They also give us the opportunity to explore the workings of our thinking mind and understand the exquisite interactions whereby our beliefs generate our thoughts, the thoughts create our feelings, and these emotions when combined with sufficient intensity are actualized as our physical experience of reality. This is the mechanism for creating our own reality which, when understood can enable us to change unproductive beliefs, release from habitual undesirable patterns of thought, and recreate for ourselves a more positive and fulfilling experience of life.

If we understand that I am the divine greater part of self, and that I preplanned my basic life scenario to learn certain lessons, and I have the power to create my own experience from moment to moment as I carry out the scenario, then what am *i* afraid of and why am *i* suffering? The answer of course is that only the *i* of us that is manifesting in the world of form is subject to suffering.

This increased awareness helps to defuse the situation and lessen it's seriousness. Even the little *i* has personal power and free will and choice, once it has gained awareness to use intellectual insights and experiential awareness to dramatically reduce suffering. As always, our point of power is both personal and immediate. What we need is to match these insights with practical methods of re-directed thought and behavior to alleviate our suffering, and give ourselves the added skills to change our personal lives and our world for the better.

PART V
HOW DO WE CHANGE OURSELVES FOR THE BETTER?

The finished man of the world must eat of
every apple once.
> —*RALPH WALDO EMERSON*
> *(Conduct of Life)*

Make wisdom your provision for the journey
from youth to old age, for it is a more
certain support than all other possessions.
> —*BIAS*

I am glad to learn, in order that I may teach.
> —*SENECA (Ad Lucilium)*

PART V
HOW DO WE CHANGE OURSELVES FOR THE BETTER?

METHODS TO CROSS THE BRIDGE BETWEEN INTELLECT AND EXPERIENCE

This is the largest section of the book, and for good reason. Now we put the laws and theories into practice. We have explored the nature of our true multi-faceted identity and existence (Part I), the source and purpose of our being here in the physical world of form (Part II), the Laws that govern our journey (Part III) and the whys and wherefores of our experiences of suffering (Part IV). Now we fit these puzzle pieces together, cross the bridge of theory and sample some practical methods to actually experience what previously may have been for us just intellectual speculation on the possibilities of bringing into our lives peace of mind and lasting happiness.

What do we know so far? We know that both individually and collectively we are all an integral part of the universal energy source, the collective conscious awareness of "All That Is," the realm of God, the Creator. We are like individual waves in the vast ocean of consciousness, grains of sand on the Universal beach. However we describe it metaphorically it all boils down to the same thing—the Universe consists of a great energy awareness of consciousness and we are individual soul sparks of that same source. The Universe is multifaceted consisting of many different levels of existence. This earth plane is just one. The earth plane is very lush and thick with physical sensations; it is the playground of form, the three dimensional amusement park we decide to visit so we can maximize our growth through preselected karmic learning tests and the everyday experiences we will encounter in this earth school. We become physical beings so that we can experience the density of the earth plane, but at the same time we, as integral parts of the big U "Universe," are multilayered beings. Our primary source connection is to Universal energy consciousness and our divinity exists at that level, while at the same time we play in the world of forms in our level of human beings. That is where we experience on a continuous basis the human qualities of mental attachments as we identify with our physical body, our thoughts, our emotions and feelings, and our continuous melodramas. As our underlying soul essence — conscious awareness journeys through this lifetime we are playing the game of walking a razor thin tightrope, trying to balance between the different worlds of our source divinity and our earth plane humanity. These multiple levels all interact and coexist simultaneously, flowing back and forth and in and out with each other in an exquisite dance of consciousness. Our game of life is danced in the creative tension between the two.

It's like we're riding a cosmic roller coaster where we go down, into the webs and shadows of our enfolding stories and dramas, kicking and screaming all the way—and then we shoot up into the light of awareness of the greater part of our beingness. Then we plunge down again into the dark depths of identification with our dramas, misconceiving them as real, and then we go back up again for air. In this constant up and down, in and out, remembering and forgetting process we keep looking "out there" into the distance for our fulfillment, satisfaction of ever changing desires, and our ultimate destination. The real Truth is that our focus is backwards. We are traveling with an incorrect map because everything we were looking for has always existed within us to begin with. We are after all connected to divinity and formed from it initially. The journey itself is it's own destination where the everyday happenings that are our experiences of life provide our vehicle to successfully navigate our way through it. Our total worldly existence takes place in each eternal now moment.

THE FORMULA FOR SUCCESSFUL REALITY CREATION

We're going to get right into the nitty-gritty of enhancing our experience of life. There is a formula to create, discreate, and re-create desired realities. It does work and anyone can successfully use it if they want to. It has four steps, all inter-connected in concept and operation, that act as the triggering mechanisms for our process of positive change. The terms are presented not in their traditional meanings, but within the context of the material already presented in prior chapters.

1) AWARENESS—This is the process of creating an opening into the multi-dimensional nature of our existence; the door through which everything else passes. This opening is a dissolving of our previous limited viewpoints and our refocusing on the greater possibilities of identity and existence. It allows us to bring them into our personal experience of consciousness:

a) Things are not always what they appear to be,
b) We are not who or what we thought we were,
c) Reality and illusion are two sides of the same coin,
d) Anything is possible, and
e) Our lives have purpose and meaning—we are the creative causes of our experience, not its effects.

In order to open we must become quieter and notice what's going on around us. Like the sign at the railroad crossing, we need to "stop, look, and listen." This is the only way we can really gain the necessary vision and knowledge of the significance of our life experiences. This is how to truly *respond* to any situation. This cultivation of conscious awareness gives us that "moment of reflection" which disconnects our circuits of robot-like reactivity. We do it paradoxically, by doing it:

a) Quieting the mind and relaxing the body through practices of meditation, concentration, and contemplation.
b) Practicing stopping, in our daily business of busyness; the art of slowing down. This requires making the time to stop

our external doing and allowing periodic resting, quietly inside.

c) Bringing quiet into our lives by—(1) Eliminating unnecessary habitual areas of noise, such as television, loud music, or family shouting matches. (2) Choosing to remove ourselves from situations of noise pollution. (3) Substituting quieter sound alternatives.

d) Really looking at our world of people, places and events on a here and now moment-to-moment basis, rather than spending our time pasting and futuring.

e) True noticing, using all our physical senses, of the exquisite details of everyday life that are usually ignored in our frenzied travels from nowhere special to nowhere in particular.

f) Listening to our inner self, from this space of quiet.

There is a kind of interior music that we can hear once we have turned down the volume of all our external distractions. It's a very personal sound; for me it is like the sweet buzzing of crickets in the night woods, combined with an almost lyric rhythm, a sort of other—worldly orchestra of unusual cosmic sounds. We all have this particular inner music which we can hear when we get quiet enough. It really is the sound of our own pulsating energy essence—the music of life!

Our suspension bridge between the theoretical concepts and our affirming life experiences is built from awareness, supported by awareness, our means of locomotion across it is awareness, and keeping our balance during the journey depends upon—yes, you guessed it—awareness! No wonder the Buddha said, "Awareness is the Lord of everything."

2) UNDERSTANDING—This is the process of *feeling* your heightened awareness; experiencing it through the interior tones of your intuitive heart rather than the logical arguments of your ego thinking mind. This is true understanding from the emotional level, where your whole beingness resonates with the conception; you absolutely *know* it is true. You totally "grok" it.

To get an idea of this feeling, reflect on any part of your life experience where you just know from your heart, without the shadow of any doubt or reservation, that something is or is not true. This is not intellectual truth, we are talking about internal feeling truth. You can even play with negatives, if necessary, just for the purpose of tasting this heightened sense of knowingness. For me, it's a sense of absolute conviction—

a) we are common source, an integral spark of the Universal energy essence which is our conscious awareness
b) our lives have meaning and purpose
c) our creative Godness is playful, compassionate and we all are loved very much
d) every person, place, thing and event in life is meaningful for learning and growth
e) the energy of conscious awareness is indestructible and eternal; it survives the death of the physical form it occupied for a particular life time.

Obviously *awareness* and *understanding* interact; the greater our awareness the higher is our level of consciousness; the more elevated our consciousness, the more access we have to the desired feeling sense of understanding; the combined raising of both greatly increases the probability that we will achieve our desired objectives. It also improves our overall focus and receptivity, similar to how the creatures of nature are sensitive to the ebb and flow of the rhythms and cycles of the seasons; a time for activity and a time for solitude. Heightened awareness also opens the flood gates of the dam of higher consciousness, and increased understanding is the great receptacle into which the waters flow. The result of this filling process is the availability of a tremendous source of potential hydro-spiritual energy to power the turbines of reality creation and skillful situational response.

3) INTENTION—This is the heating process by virtue of which the potential energy source that we generated through steps one and two can be actualized. Imagine that we have our huge pool of creative energy, but it is at rest—cool and inert. Once

it is heated to the required temperature however it becomes energy in motion; it expands, it overflows its previous limiting boundaries, and it can hotly engrave the deep outlines of the desired objective reality. Steps one and two give us the ability to analyze and appreciate our predicament, to see it in all its subtle nuances and to intuit the appropriate method of dealing with it. Steps three and four advance the energy so that the desired form of reality occurs.

Another way of looking at intention is intensity. This is how we turn on the burners for the heating process. We must really have determination, single-mindedness of purpose, and clarity of objective to generate enough heat to run the engine. Usually this occurs when we have experienced so much of the suffering aspects of a situation, we reach a point of firm resolve to do something about it. Intention is the step that precedes action because action is movement and the turbines can't move until the fuel energy is sufficiently heated.

Intensity can come from the focused thoughts of our mind in the form of will power. In this form it's like a laser beam of energy, resulting from the one pointedness of concentration. We image a narrow tunnel connecting our informed awareness to the desired objective and direct the energy of consciousness through it. It combusts with tremendous pressure producing the desired powerful effect. Mental techniques of creative visualization, positive imagining and waking dreams are effective tools to promote the process.

Heart power can serve the same purpose. It is the use of the emotional/feeling heat in any situation to provide the catalyst to constructively change it or transcend it entirely. This is a situational use where the built in intensity of a disturbing event provides the ready made power source to uplevel it in the desired way. When our emotional buttons are pushed, the fire for this chemical process is right there in front, available for immediate use. All we need do is re-channel ourselves from the reactive, negative-closed mode (where we initially feel the heat) to the responsive, positive-open mode — using our awareness and understanding skills. We then are ready to constructively use this energy to complete the equation with overt action on the physical plane.

4) ACTION — This is the fourth and final step of this process, where we complete the painting of the previously outlined black and white possibilities by painting them in living colors. We accomplish this by moving forward with consistent action in our world of physical form. All the preparations have been made and all that's left is for us to resolutely put them into forward gear.

The movement inherent in physical action not only feeds upon the energy source created and primed in steps one through three; it also increases it in quality and intensity, thus augmenting the process. This is the phase where you go out and "do the deed" — you attend the health club, change your diet, join a twelve step program, enroll in the school, ask her/him for the date, interview for the job, etc. It is the *affirmation* of doing that changes an alternate probability forged by the awareness-understanding-intention link into the final fact of physical actualization. It is the closing of the electrical circuit between the invisible and physical worlds by virtue of which the desired reality emerges from its birth channel into the light of day.

THE KEYS TO NON-SUFFERING

We've already explored the nature of suffering in Part IV. Our practical interest, of course, is how we can have a more consistent and lasting personal experience of its other side—happiness (also known as joy, rapture, ecstacy, peace of mind, and contentment). The most pragmatic explanation I've heard was given by Ken Keyes in his "Living Love" workshops; "Change what's changeable if you can, and if you can't, accept the Unacceptable." It follows the well known 12 step prayer:

"God grant me the serenity to accept what I cannot change,
The courage to change what I can, and
The wisdom to know the difference between them."

We again, like the prior section on reality creation, have a workable formula for success.

The key is *acceptance* — the mental quality of allowing something to be the way it is even if it doesn't satisfy our expectations or meet our desired model of reality. When we consider something to be unacceptable, as not being the result we wanted, we only will suffer if we are *emotionally* demanding it to be different. This is where our thinking mind entraps us in the drama of our ABC suffering scenario we talked about in Part IV:

a) We want something to happen but it doesn't (grasping)
b) We don't want something to happen, but it does (resisting)
c) We want something to happen, it does, but that something doesn't last (clinging)

Therefore, the broadest statement for happiness is "Accept What Is"!

Within this maxim is a trio of factors that are linked together in theory and operation:

1) ACCEPT LIFE AS IT COMES—Not everything we encounter in life is to our liking. The broad spectrum of our experiences of people, places and things does not always fall within the proscribed boundaries of our personalities, mental models and societal desires. If it did we would be bored to tears and never learn the important karmic lessons we've scripted. We are sparks of energy, part of the Universal creative force and the most singular characteristic of creativity is constant challenge to expand it's limits. This requires events that push and pull us, driving us to maintain the delicate balance of creative tension that allows us to function in the world of forms as an effective organism.

Since the primary characteristic of the world of form is constant change, our greatest constructive challenge becomes the flexibility to continuously open ourselves so that we flow with these changes of life while staying grounded enough to effectively function in our everyday world. This is what true acceptance is all about. Our lives are a continuous procession of experiences that both test us and give us the opportunity to achieve this understanding.

"Fight or flight" programming is deeply layered into our personality, explaining the hair-trigger nature of our emotionality when we are faced with a provoking situation. Almost before we are conscious of our actions, we have begun another round of attack and counter-attack, increasing our separateness, closing our heart and getting lost in our drama. How do we learn to suspend this immediate inclination to react? Gradually, we practice the qualities of noticing awareness (STOP, LOOK and LISTEN). But the fires of emotion that burn us are painful and we have been trained since infancy to avoid the flames. What do we do? We try to sit right in the middle of the inferno, "cooking" — so that the thickly encrusted layers begin to melt and flow away. Sometimes we may simmer, other times we may be broiling; however, these energies are always in motion, and they will change of their own accord, if we can have the patience to let them. By allowing this process to unfold we will gain the ability and insights to grow into the spaciousness and clarity that characterizes true responsiveness.

2) LET GO OF UNFULFILLED DESIRES—Imagine trying to stop the flow of a cascading waterfall and becoming upset when you do not succeed. Isn't it better to acquiesce and enjoy this inevitable movement of nature? Letting go requires us to give up our emotional struggle to reach our goal by "pushing the river" of our life experiences. Instead, we must stop resisting its flow, and allow it to carry us along, in it's own way of things. When we release from our attachments to a required happening of this or that, we can reclaim the tremendous energy we had invested in our futile efforts. This allows both an emotional and physical resurgence that cleans out our blockages and restores our sense of internal balance. It puts us back in harmony. We notice it as we are energized again, we feel so much lighter, and we are open and flexible. As Dick Sutphen says, "Out of this acceptance (in letting go) comes conscious detachment; the ability to enjoy all the positive aspects of life, but to allow the negativity to flow through you without resistance and without affecting you."

This energy surge also gives us the mental clarity to see that there are alternative practical ways to probably accomplish indirectly what we are unable to do directly. When we are emotionally upset, lost in our dramas of wanting a particular happening, we live in an immature mindset of either-or programming. Observe a young child in a shopping center who has a tantrum when Mommy doesn't provide the instant gratification of buying a desired ice cream cone. The youngster is totally consumed emotionally, seemingly heartbroken, until an alternative is presented such as the idea of stopping at a restaurant after shopping is completed, or having a box of popcorn at the mall movie theatre once Mommy is allowed to finish her task. Suddenly, the tears vanish and the sun shines again. We regress to this childish mind whenever we become sucked in by similar circumstances. As we lighten, open, and calm ourselves, we regain the vision to see the *and's* of every situation—how we can probably still accomplish most, if not all, of what we wanted to do and get most of what we desired.

3) TRUST THAT WHAT'S BEST WILL FIND YOU—This is absolutely essential for the trio of happiness methods to be effective. When we *let go* of what we previously had been stubbornly holding onto, and *accept* the present circumstances, we are making a leap of faith. We are facing head on our basic survival fears of the unknown caused by the unpredictability of sudden change. In the words of Appollinaire,

> "Come to the edge, he said.
> They said, We are afraid.
> Come to the edge, he said,
> They came.
> He pushed them—and they flew."

In order to symbolically jump off the cliff by ending our resistance to what is, we need the comfort of knowing we have a safety net. This reassurance is provided us on an almost daily basis by the Universe, if we look closely enough, in the form of synchronous events that provide us with proofs that we are in harmony with it, and that all is well. These moments of "meaningful coincidences" are everywhere if we have the conscious awareness to notice them. You wake up wondering about the whereabouts of your income tax refund and it arrives in that morning's mail; or more graphically representative—you've been struggling for months with the decision whether or not to quit your unfulfilling job which you have kept solely for financial reasons, you finally "bite the bullet" and tender your resignation—not knowing where your next mortgage payment will come from, and that afternoon you receive notification that your rich Uncle Fred died and your inheritance from his estate will make you financially independent.

This type of trusting concerns the intuitive and feeling level of true understanding, where we know that a particular event is just another part of the exquisite perfection of the Universal plan. Even if we've reached a dead end in a particular situation everything will work out for the best, if we *accept, let go,* and *trust* that it will. This is the firm foundation that we can use as the spring board for practicing our dives of acceptance. As Sibyl Johnston says, "God wears many disguises. He surrounds himself with paradox to keep us from approaching him in any way but by faith."

Sometimes our faith is shaken and we experience what is commonly known as "The dark night of the soul," when it appears all is lost. At these times especially, when we think we have nothing to hold onto, we must keep the faith, for "Never are we nearer to the light, than at the darkest time of night."

Like all other aspects of the Game of life, the more we practice these formulas for success and happiness in our daily lives, the more skillful we become. By doing so, less practice is necessary as they become integrated as a regular part of our daily lives.

Eventually as we offer less and less resistance to the natural flows of Universal energy in giving us what we really want, we don't have to do anything other than wait patiently and confidently as we allow it all to unfold for us in its natural way. We don't have to grab for it; rather we can just sit back and receive it.

To experience the feeling of Trusting the Universe, you have to be prepared to also let go of *being in control* of every situation (Some things in life are just not changeable, or the emotional investment is just too great) by habitually molding them to meet your approval.

For example, if I am in an addictive romantic relationship where I believe my well-being is dependent upon my partner not looking at attractive men, my emotional struggle centers around trying to control her so she will meet my model of required behavior. If I am a person who is neat and orderly, and my roommate is sloppy and disorganized, my efforts to change the "Oscar Madison" character into a "Felix Unger" odd couple stereotype are really control issues, because I am not accepting or allowing what is.

Being out of control is a feeling we usually avoid at all costs. It is fear with a capital F; a doorway to the great unknown that we don't want to enter. We believe we must constantly monitor and manipulate our people—place—thing events because they will not meet our proscribed models unless we do so.

Those perceptual models are the problem. We don't trust we can leave anything to chance. We equate such non-action with a foolish disregard of our perceived need to force—to make things happen. Why? Because we lack trust that they will happen anyway, in their own way of things, if we just allow the events of our lives to properly unfold.

These efforts of ours to push the river of life are futile. It will take us where it chooses to go anyway, meandering here and there, and ultimately reaching the same destination. We can travel first class, enjoying all the sights, or lowest class where we kick and scream, resisting all the way.

The physical and emotional energy we spend trying to control our lives is tremendous. Being constantly on red alert is exhausting work. All the tapes of programmed behavior that run us—the do's and don'ts, shoulds and ought to's of our lives, must be powered constantly because they are not an integral part of our inherent being. Rather, they have been artificially engrafted upon us in the layering process. Once the energy that keeps the layers in place is withdrawn, they begin to fall away.

There are safe and interesting ways to practice experiencing the physical sensations and emotional feelings of being out of control, allowing it all to just happen, and trusting that all is well.

Visit an amusement park and go on the rides you were previously afraid to sample. Don't resist your fear of being out of control; instead, just notice its ebb and flow and listen to the fear tapes playing in your mind.

Risk being more adventurous. Do things never experienced before. For me, this would include getting on horseback, ascending in a hot air balloon, taking a helicopter ride, wind surfing, and ice skating. All instances of feeling out of control. Make your own list of "scary" things to do and have fun adding to your inventory of personal adventures.

As we become aware of the outlines of our fears, we can gently approach them and test their limits. As we do, they evaporate and are transformed into a sense of personal accomplishment and a deeper understanding of how the fact of living requires trust on a moment to moment basis. It's impossible to be in control all the time and, more importantly, it's not necessary.

TURNING OUR WORLD OUTSIDE-IN

We understand at this point of our journey that our emotional experience of life as heaven or hell depends totally upon our *attitudinal perspective*. Is our outer world the reflection of our inner state of consciousness, or vice-versa? When our human personality commences it's layering process and starts to develop and focus within the world of form, it constantly seeks fulfillment from the outside world. It focuses it's lens of perception "out there" consistent with the various "teachings" that are received from parents, peers, and society in general. Our underlying pure energy essence conscious awareness is like a flashlight beam that illuminates everything. When we started our physical lifetime this inner light was shining everywhere and our focus of awareness was internal. We knew intuitively that, like the Bible tells us, "The kingdom of God lies within."

We also had the encoded wisdom that all our physical experiences would be created first and foremost in our interior environment through our belief systems, moved forward through our resultant thoughts and emotions, and then be realized as physical experiences in the outer world. This critical understanding and initial focus inside-out enabled us to maintain our connectedness with our soul essence and remain in harmonious balance as we walked the divinity-humanity tightrope. However, with the seductiveness of the world viewed through our physical senses, the bombardment of negative belief systems upon us, and the completing of the layering process—the inner light dimmed. As our physical incarnation developed it was as if we took the flashlight from within and reversed its beam to face outside of us. That is where our internal awareness darkened and we initially lost our way. So we staggered around in our own living room looking for the way back home.

As we lost our way in the layering process and went from the wisdom, sense of well-being, calm and loving compassionateness of "inside" and headed "outside," this erroneous focus was continuously reinforced in our western society. God was "out there" according to our Judeo-Christian religious leaders, and the tangible material evidence of our success and happiness was also external to us in the form of striving to achieve the best career, the most prestigious job, the shiniest car, the biggest house, and the most glamorous relationship. We've all taken that trip, and as we look back on it now with a degree of detachment from the vantage point of our higher self, we can impartially observe that we really had the gear shift in reverse all the way. It was like we were riding a horse facing backwards and wondering why it was taking so long to get to our destination.

Taking this wrong fork in the road also caused us to misperceive the spiritual purpose of our life's journey. Once we gained a certain degree of awareness so that we at least understood that our identity and existence is multileveled, we still had a tendency to retain our western orientation of achieving goals and we therefore thought that the purpose of our journey was also "out there." We forgot that on this pathless path the goal of our journey is the everyday experience itself, as a means of learning our necessary lessons for awakening, rather than any ultimate spiritual destination. From this standpoint we are not limited, flawed beings, tainted with original sin who are trying to find a God who lives on the mountain top. If this is our belief, we are always craning our necks, gazing upwards at the peak that seems always out of reach. We also have to undergo a hard and dangerous climb upward in order to reach the gates where, even after our arrival, there is no guarantee that we will be admitted to this imagined exterior paradise because a judgmental determination must be made as to whether we are worthy.

This perspective is an inverted pyramid where we are the tiny triangular point at the bottom and we have to somehow traverse the sheer walls up to the broad base of enlightenment that exists at the top. If, however, we changed our focus *outside-in* and turned this pyramid upside-down, we would have an easy and exhilarating slide back down to our greater beingness whose broad base we were standing on all along while we were gazing upward.

In our usual drama of climbing up the mountain we struggle to do it and make it happen the difficult external way we have been conditioned to believe it must be. Each new learning experience seems to cut a new foothold for us as we laboriously struggle to dig in our pitons and head up the slope. But we are only treating symptoms, where we keep applying successive layers of band-aids over our wounds to temporarily stop the bleeding so we can move on, even though the required healing has not yet occurred to the infected layers lying underneath. By the time we reach the top (if we ever do), we are exhausted, sore, still suffering from the same layered problems, and not much better off than when we started.

If however we go *outside-in,* each successive learning experience completely removes another layer of our "stuff," so that eventually all the layers fall away and we are completely healed inside with a new skin of heightened awareness and raised consciousness. This is true preventive medicine; gain without so much pain.

It's really all a matter of adjusting the focus of our awareness. If we look through the wrong end of a telescope we can't see clearly. The images are cloudy and unclear. The same holds true if our flashlight is pointed in the wrong direction. It's difficult to get our bearings and we misperceive the outside as what we've really been seeking inside. While we were trying so hard to find ourselves outside, we were looking everywhere except the most obvious place which was of course inside.

In our external focus we are ever striving to grab the brass ring of happiness that seems just beyond our grasp. However, in truth we have never let it go to begin with. This fact is forgotten as we are deluded in our personal belief systems of lack and limitation and suffering. This is the conspiracy of unconsciousness that is the layering process which causes us to lose our way and clouds our lens of perception. We are not human beings trying to find our divinity; rather, we are divine beings using our humanity as our vehicle to grow.

So when we have this awareness and understanding we can move these theoretical abstractions into real life experiences that affirm them as valid for everyday use. The method is simple—use your physical senses to evaluate whether your focus is overly external. In any people, place, or thing phenomena which raises perplexing questions in the mind, uncomfortable sensations in the body or seems to be otherwise disharmonious, check yourself out and see where you are shining your flashlight of awareness. Is this

an instance of reaching or grasping? Is there avoidance or resistance? Is there a clinging to the past or an expectation of the
future, rather than a welcoming acceptance of the here and now
experience?

This self monitoring process may first appear to be a lot of hard
work. It really isn't once you get used to it and put it on automatic
pilot. There develops within your heightened awareness the necessary processing which uses your thinking mind as an ally rather
than an adversary. It can help you, through your brain bio computer, to stay on the path through your journey of awakening. It's
as if you have placed a homing mechanism on your vehicle and
whenever you start to drift off the road it alerts you to the fact so
that you can see what's going on, understand it, and then make the
necessary adjustments to straighten out your course. You use your
rational sensory mechanisms to help detach from getting lost in
them.

After we employ this *outside-in* approach of personal awareness
for awhile, and it becomes more automatic, it is not even necessary
when we stray off our course to understand the WHY of it. Rather,
we can just notice, accept and acknowledge, make the necessary
adjustments and continue on our way. This is a much more practical self healing approach then beating on ourself for making a
mistake or spinning around endlessly in whys and wherefores of
our past dramas. We finally realize that we don't *find* happiness
"out there." It always exists within us as a state of being. It
emerges of it's own accord when we allow it to, by quieting our
chattering mind and opening our heart to reveal it.

THE POWER OF BELIEFS

The process of natural self adjustment that we talked about in
refocusing our lens of awareness *outside-in* is also helpful in becoming aware of our internal belief systems and habitual patterns
of thought. They are creating our exterior experience of physical
reality, and not vice-versa. They run the tapes of programming
that move us through life. We experience what we believe we will
experience. Our correct focus changes from, "I'll believe it when I
see it," to "I'll see it when (and if) I believe it."

The laws of energy and attraction that we examined previously are the basis for this concept of reality creation. We and our Universe are continuously flowing and changing energy. The solar system within which we reside is energy as is everything else within "All That Is". In our evolutionary processes, our consciousness seeks higher and higher frequencies in the universal process of creativity. And it constantly creates matter. The physical Universe arises from consciousness — not the other way around. Einstein's theory opened scientific minds to consider the probabilities that our Universe and everything within it is multilayered, occupying different levels of relative reality. They act like filters, tuning us in to a certain range of experience. If we change our beliefs, our experience also changes. The table that appears solid to the perception of physical senses and the labeling of our conceptual mind as "solid table" is on another equally valid level of reality merely a phenomena of atoms and molecules interacting together with the necessary speed and vibrational frequencies to present the illusion of solidity. When speaking of this invisible world of alternative reality, Einstein said, "I did not reach my understanding of the Universe through my thinking mind."

As we noted before, the thinking mind is a subtle aspect of our physical world of form. The larger invisible world from which form is created is meta to it. Current scientific acceptance of these concepts continues with the expanding ideas of quantum physics, where the world of science is completing the circle of understanding back to the mystical/spiritual theories of identity and existence upon which our exploration of higher consciousness is based.

The fields of spiraling energy which surround us include the magnetic auric field around our physical body. As such it attracts to itself whatever fulfills the beliefs embedded in it. The beliefs generate thought forms which are actualized by intense force of will into our physical experience of reality. Therefore one can hardly blame anything exterior as causing an undesirable physical reality. If we examine our belief systems we can notice our patterns of thought which have generated our feelings and emotions and which in turn, with the energy catalyst of intentional action, materialize our physical reality. The underlying greater part of ourselves, our divinity, is just doing what God's do, the Creator continuously manifests its "belief systems" into physical reality.

Even the past is a product of a belief system which we are creating from moment to moment. It's an idea that exists in present time, created to give us supporting evidence of what we believed at one time to be true. We act the way we do because of our present beliefs, not our past actions. They were just the reflection of our belief systems that existed at that time. As our beliefs change, so does our experience of reality.

We need to locate and identify belief systems which no longer serve us and from which we are creating for ourselves disharmony and personal suffering. Once we locate them we need to modify them or totally dismantle them and recreate a more desireable experience of reality through the more positive energies of upleveling them. If we can learn to do this we can become true time travelers.

With this heightened awareness and the skillful techniques to implement it, we can explore numerous belief systems to learn our lessons of awareness during our visit to the earth plane. After we've explored a particular belief, like a passport, we can discreate it and walk away from it without getting trapped within the personal dramas which are it's experiential product. We can then create another belief system to explore and so on and so on. While we explore a belief system we can, on our humanity level, become the somebody of that particular drama and wear their identity costume. We can discreate it when we're done, take off the costume and again assume our non-identity as we prepare to repeat the process. In maximizing it's effectiveness for our personal growth we can have the detached perspective where we are the explorer and not our exploration. This would then be our constant flowing process of non-time/time travel.

A helpful hint for identifying the belief culprit in your current emotional drama is becoming aware of the mental chatter and then zeroing in on what your mind is demanding. The trigger device is the "I want" of the dialogue. As we continue to use our *outside-in* approach, we also note that the *opposite* of the fulfilling —granting of the demand is in fact the belief system which you are seeking. For example in the area of relationship, your mind may be demanding, "I want her to fall in love with me." The underlying belief may be its opposite, "I'm not lovable." Your mind may say "I want her to find me attractive." You may really believe "I'm not attractive." Your mind may suggest, "I want her with me so I will be happy." The underlying belief may be that "My happiness is dependant on others." You get the idea, the possible examples are endless; the point is that this will help you in your investigative process. Remember the unpleasant bodily and emotional sensations you experience are a proof that you have scored a bulls eye. You have zeroed in on an area where your mind is holding dysfunctional beliefs and acting them out to your discomfort.

Notice your personal characteristics that occur when someone or something pushes your buttons. There is tightness, constriction, shortness of breath, heaviness, blocked energy and lack of humor. This state of seriousness, closure, withdrawal into a shell and the combination of a distracted chattering mind and a heavy closed heart is prima facie evidence that you are on the right trail of locating the offending belief. Also be aware of the opposite physical bodily sensations and emotions which are your natural state of being. When things are going well and we are experiencing what we label as happiness, contentment, and a sense of well-being—we are furnished with unmistakable proof again that at that moment there are no events that are triggering any of our negative programming. Our mental models are being met and none of the A, B, C, mind scenarios that result in suffering are being played out.

BECOME A PRIVATE EYE

We must be aware of clues that suggest to us a particular experience of reality is detrimental to our growth and thus needs to be examined, for possible change or discreation.

This examination is a part of the handyman process that is an integral part of our expanding awareness. We locate areas that need fixing and then we do the necessary remedial work in as skillful a manner as we can. This is a part of our growing list of various types of coping mechanisms to maximize our growth.

In the workshops, lectures and New Age discussion groups that I am involved in I always ask the participants about their personal bag of coping tricks. Often they have been unaware that there were specific things that they did in order to deal with stress, emotional upsets, heavy personal dramas or any other events which pushed their emotional buttons. Whether we have automatic systems of corrective behavior or those accomplished with intentionality, they still help us to maintain our balance, stay afloat, and continue on course.

We must seek to become attentive to our internal clues so that we can solve the external mysteries that are causing our physical experience of reality to be unpleasant, unfulfilling or unbalanced.

1) Quiet down and notice what is going on in your personal drama. What are the physical sensations of discomfort or disharmony that are being experienced? Scan your physical body to locate pain, tightness, general discomfort, or other physiological symptoms of upset. In order to do this you must STOP whatever you're doing at the moment and just suspend all physical activity while *scoping* your entire body. Write down the sensations as clues and the areas of the body which are affected. This not only tells you what is happening in connection with this present situation, but also helps you to understand your personal physiological signals of upset.

2) Scan your emotional body in the same way. What is your mind saying? This internal dialogue will, at the time of an upsetting occurrence, be loudly holding center stage in your movie of consciousness—it will probably involve aspects of judging, criticizing, comparing, labeling, and many other of the mind —games which are familiar to you. The key is to heighten your awareness so that you notice the dialogue and record it. It will lead you directly to the beliefs that generate it and their physical manifestations which are acted out.

This dialogue is your internal verbalization of your belief systems—in computer terms it's your programming. Since one of the characteristics of thinking mind is it's reactivity, we are assured that it's dialogue at these moments is a valid indicator of the non-productive tape that has been playing in your head time and again when similar situations arose. Previously, you were not sufficiently aware to understand the problem and therefore unable to correctly focus your attention on where and how to fix it. Write it down, in summary form, in your detective journal.

3) Once you become used to this *outside-in* approach to process current upsetting situations, you can gain additional clarity into habitual patterns of reoccurring past situations. You should therefore use the instance of any button-pushing incident to not only explore it in order to get back to it's root belief causation, but also use this opportunity of stopping and noticing to focus your mind on any other vivid instances in the past where you have experienced similar symptoms of suffering as a result of people, place, or event circumstances. Write down the past factual event and the circumstances under which it took place, just as if you were writing a newspaper account of the incident. Be a good reporter and seek to eliminate extraneous detail; just the facts please. After you have written down the narrative, close your eyes and attempt to visualize it in living color and relive it from a physical and emotional standpoint. Allow all the sensations and emotions to come to life and flow through you, while at the same time reassuring yourself that this is only a movie that you are viewing as an interested observer in an experiment that you are conducting. You are not your drama, you are just watching it as you play the role of detective. Write down the physical sensations, the emotional feelings, and the mind dialogue in the same manner as you did for the present event processing. Are different belief systems uncovered, or is there a repetitive pattern of sensation—feeling—belief which occurs?

4) There may have been major dramas in your life which you have viewed up until now as tragedies, intense melodrama, negative experiences, or just plain bad luck. In any case follow the same processing as we have been doing by stopping what you're doing, and *quieting down in a space of privacy and silence.* Look at the factual movie again and relive it's intensity from the perspective of the detached observer. Write down the factual narrative and the physical — emotional — belief checklist. Repeat this same process for as many of your major life dramas as you can recall. What probably will happen as you start to think about them is they will all flood back into your recollection with detail and intensity. If they may appear jumbled and disconnected, just continue to quietly notice and attempt to view them in a sequential manner. This

will help organize them in a time sequence for your investigation.

5) Ego Traps—The most difficult part of the process, once you make up your mind to do it and allow yourself to experience and re-experience unpleasant events, is actually locating a triggering belief. They usually are very well hidden behind many layers of surrounding programming. You may think you have located a belief but it may only be a camouflage idea of the ego that insulates the underlying core belief from being discovered. Remember the ego thinking mind is very clever and will do whatever it can to throw the detective off the trail of solving the mystery.

The ego is an excellent games player once it perceives that someone is breathing hotly on it's trail. It will also try to convince you that the work you're doing uncovering your negative belief systems is too hard, it's weird to go into these matters, the pain from dredging up old memories will be too intense, or why not go to your nearest bar and have a nice cold one? It knows how helpful this linking process is where you connect your suffering with the belief systems that have caused it, and it tries to protect it's continued role of dictator through familiar repetitive patterns of behavior.

There is also another paradoxical trap which your ego thinking mind will seize on. The mere process of refocusing your perception and going inside to analyze the connections between beliefs and actualized physical events *will* make you feel better. The process releases the clingings and attachments of your mind in those areas of blockage that have been producing the painful sensations. By increasing your awareness you are detaching from your drama, disidentifying with your personal movie, and this elevated perspective loosens the areas of mental clinging. You clear the mental blockages and release the trapped energy that produced the sensations of suffering. It is a catharsis process and it usually produces immediate release which you feel as relief from your discomfort.

When this occurs your ego mind will be telling you that everything is now OK and you don't have to pursue this introspective process any longer. You can go back about your everyday business and refocus on your exterior world of "reality." When this occurs, by all means welcome the initial relief as an affirmation that you are on the right track but don't turn back just yet. In fact this unmistakable clue that you are now focusing in the right direction should give you additional determination and energy to move further forward and complete all the necessary steps in your attempts to locate the hidden negative beliefs that have been running your life.

6) The Dis-A-Scope — Think of your concentrated focus as a flashlight beam that you are shining inside in every widening circles. You are looking for the areas where you are dis-enchanted, dis-connected, dis-turbed, and dis-tracted. The obvious key in uncovering the negative or unproductive beliefs in this linkage of causation is awareness, awareness and yet even more expanded awareness. It is both the goal and the method for achieving success.

We can also know when our awareness has been sufficiently heightened, when we are balanced, centered, and in the natural harmonious flow of universal energy. As we stop, look, and listen to all our mind-body processes, we can experience what it feels like to rest within the greater part of our conscious awareness. This is the peak experience; a state of being that is hard to describe because it is beyond our verbal world of form. But W. P. Kinsella in "Shoeless Joe" (the basis for the movie "Field of Dreams") comes close:

"There is a magic about it," I say. "You have to be there to feel the magic."

"What is this magic you keep talking about?"

"It's the place and the time. The right place and right time. Iowa is the right place, and the time is right, too—a time when all the cosmic tumblers have clicked into place and the universe opens up for a few seconds, or hours, and shows you what is possible."

"And what do you see? What do you feel?"

"Your mind stops, hangs suspended like a glowing Chinese lantern, and you feel a sensation of wonder, of awe, a tingling, a shortness of breath . . ."

"And then?"

"And then you not only see, but hear, and smell, and taste, and touch whatever is closest to your heart's desire. Your secret dreams that grow over the years like apple seeds sown in your belly, grow up through you in leafy wonder and finally sprout through your skin, gentle and soft and wondrous, and they breathe and have a life of their own . . ."

"You've done this?"

"A time or two."

"Is it always the same?"

"It is and it isn't."

ELIMINATING OUR FEARS

The most characteristic of our unproductive beliefs are of course their negativity. All are essentially fear based with negative emotional subsystems. Our A, B, C, scenario of suffering with it's reaching and grasping, resisting and avoiding and clinging to the status quo is all a product of fear.

The next step after we have located the negative belief systems that are the triggering devices for our unhappiness, and before discreating them in whole or part, is understanding the origin of this negative input and where in our life it is being reinforced. As long as the supply of it's energy continues it will be very difficult to successfully reprogram ourselves. As long as garbage is coming into our computer it will produce garbage. It's like putting clothes out to dry underneath a leaking faucet and wondering why they're always wet when we put them on. Again we need the awareness of what's causing the problem. The key is our fears; our free-floating anxiety.

Where do we get all this negative fear programing? As we discussed when we looked at the layering process in our personality development, our earliest teachers are our parents, school instructors, family surroundings, and our peers. Since they all carry negative programming (it is so widespread that we all have it whether we realize it or not) it is inevitable that our earliest mental tapes are duplications of theirs. They may have inherited this input from the same sources, and so on back down the line, with each succeeding generation adding new layers of fear based reactivity as the society in which the individuals were living became more complex, faster paced, and with a greater sense of individual isolation, anxiety, and helplessness.

This is further reinforced on a daily basis by the media; the irony being that as our society has "advanced" in the 20th century with the industrial revolution, the advances of scientific technology, and the era of instant world wide communications—there has been a corresponding dramatic increase in the reporting of gloom and doom on a regular basis.

You have proof of this everywhere. Turn on your TV news or pick up your newspaper. The lead stories and articles bombard you with negativity and fear. The picture that's painted for you is of a world gone mad where danger lurks at every corner and your physical health and emotional well being is constantly under attack. Since your attention is also primarily focused outside of yourself on this external world, you also receive confirming evidence on a daily basis that this negative perspective must be true. It is a vicious cycle of self-fulfilling beliefs and events. Further reinforcement occurs with the "entertainment" that is offered by the media where gratuitous violence is shown continuously as a natural by-product of our contemporary life.

A lot of this negative fear based programming has also come from our own personal experiences. As we act out our personal dramas, to the extent that we carry negative fear based beliefs with us we have actualized them by experiences of pain and suffering in our daily lives. The more of this negativity that has occurred, the more we are convinced that life is meant to be this way, and of course the more we believe it—the more we create it. It is also self generating and perpetuates the cycle.

We must find our fears so that they can be consciously examined, processed and eliminated. Fear is insidious; it is constriction and darkness. It is removed by the light of awareness and the opening of acceptance. A good process is to list your fears in your journal, in the safety and privacy of your quiet physical space. As you list them, think about what past or future experiences, either real or imagined, they bring to the movie screen of your consciousness. Allow the physical sensations and emotional—feelings states to take hold of you so that you can gain a deeper awareness of the fears. Remember you are perfectly safe, and this is a positive therapeutic process. As it unfolds, much of the fear energy vanishes by itself since it was the repression of these events and your mental models of their significance which was what made them so powerful.

Fears can be further diffused by talking about them to others who can listen consciously, without judgement or criticism. This "safe harbor" of communication can be available through conversations with a close friend, participation in support groups, counselling therapy, and other similar forums. The effect is always beneficial because talking about the hidden fears and bringing them out into the light often eliminates them. If the fears retain their power even after individual and group communication, through the tell-tale clues of physical and emotional sensation that we have discussed, we can move forward to "safely" experience them by imagining and visualization. Our subconscious mind does not differentiate between real and imagined experiences. Therefore it is a perfect testing ground for us to pretend to live the fearful situation, allow the spectrum of all its sensations, and then diffuse it by visualizing a successful conclusion. In so doing we uplevel this fear into just another one of our many experiences.

For example, let's say the fear is the experience of riding a horse. Perhaps it arose from a childhood experience of being thrown off a pony, bitten by a similar animal, or seeing a movie in which a horse became an evil demon that attacked children. Or there may be no lifetime derivation for the fear; perhaps it is a karmic situation, where a prior lifetime was ended in a battle where you were trampled by a horse. The derivation doesn't matter as long as we can clearly identify the fear.

After remembering it and re-experiencing it physically and emotionally in your own quiet safe space, see yourself very confidently approaching a gentle, loving horse. Smile as you take hold of the bridle and saddle, put your left foot into the stirrup and lift yourself onto the horse. You sit atop the animal, very comfortable and relaxed, as you lightly hold the reins. You reach forward and stroke the mane, and feel its smooth solidity. You begin to trot and gradually accelerate into a gallop as you go around the exercise ring. You see yourself as an accomplished equestrian, dressed in the traditional red riding suit and shiny black boots. There is a crowd of people watching you, cheering, as you and your horse gracefully move around the circle, executing various riding maneuvers. In this imaging process you can draw upon all your personal background from movies, books and other sources to provide you with a mental picture of an enjoyable, safe and positive experience. It then becomes a permanent part of your memory bank, notwithstanding the fact that it was an imagined adventure. The more intensely you experience it in your visualizing process, the more "real" it will be for you.

It has been suggested by many sources that letting go of our fears may be the most important aspect of our internal processing. There is a consensus among spiritual texts, mystics, philosophers, and contemporary New Age authorities that the purpose of physical existence is learning to live a life of unconditional love. This is interpreted to mean the reconnection with that greater part of ourselves where everything is related and we live out our inherent divinity through learning how to accept all that is. Jerry Jampolsky suggests, "(Unconditional) love is letting go of fear." Dick Sutphen says we are here to learn how to, "Let go of fear and express unconditional love." The elimination of our fears serves to quiet our mind and open our hearts, thus leading us toward our goals of awakening.

RE-CREATING A BETTER REALITY

How then can we discreate belief systems that don't work for us, lessen the grip of fear-based emotions on our lives, and re-create more fulfilling and happy experiences?

1. We need to input new and better information into our personal bio-computer. This inserting of new tapes to replace the old ones that don't serve us can be started immediately so that we can taste the fruit of a more positive point of view and a better balancing out of the societal negativity that surrounds us.

No one is forcing you to read the newspaper headlines or input the electronic media news. Your ego mind may say, "Well if I don't do that how will I be well-informed? I won't know what's going on in the world." That's a bunch of ego denial. When you need to know about something it will come to your attention in everyday conversation, in a bulletin that comes through the media or in other synchronous ways. In the meantime the repetitive drone of negativity is just churning out variations on the same theme and you do not have to be in it's path.

Look at how your newspaper is set up. The front pages and the bold type blares out stories of fear and negativity. If there is any reporting of "human interest" stories that are positive or uplifting they are presented as specialty items on the back pages. Look at the formatting of your TV news. The lead stories are like the Four Horsemen of the Apocalypse—famine, pestilence, war and death. This continues throughout the newscast, assisted by instant replays of plane crashes and disaster pictures in living color. The final news story however is "a light one" where the anchor person focuses on a human interest story that may combine kindness, caring, compassion and humor. After it is related to the audience, the commentator smiles broadly and as the engineers roll the credits the words are heard, "Well, good night and may you have a better tomorrow."

For years I've been trying to get local media people interested in presenting a segment called, "The good news." The wire services have many stories that fall within this category, however they are not deemed newsworthy by the media and, like the old adage which is the measuring stick for reporting —"Will it sell newspapers?" the decision is made not to carry these items. I'll keep trying and one of these days perhaps this area of media belief systems will change and create the more desirable reality of positive communication.

2. We can change how we view and perceive situations. Seth of the Jane Robert's channelings says, "You create your own reality through your thoughts, ideas and beliefs . . . your point of power is always in the present." This is very true. No person or event "out there" is doing it to us. When we go *outside-in* and increase our awareness of how we are doing it to ourself all the time, we can be reassured that we have always been the captain of our ship, we are the masters of our own destiny. We have full self responsibility for everything that happens in our life. We can turn things around for the better in the

next now moment, by seeing the hidden positive benefits of personal growth and learning in all experiences.

3. There are other ways we can turn a new page and change our lives, and our belief systems for the better. We can create a new, more positive belief by first initiating consistent affirming action on the physical level of experience and then turning *outside-in* to link consistent beliefs.

For example, let's say that one of the areas in which we need to change beliefs is that of personal relationship. This is not unusual by the way. Probably the most pervasive negative belief in our society is in the area of self esteem and personal worthiness. Such beliefs attract feelings of insecurity, shyness, aversion and anxiety; resulting in affirming experiences of relationship deprivation. This is further reinforced in our society by our relationship pattern of looking "out there" for Mr. or Ms. Right. We believe they will complete us and fill the gap in our lives, and make us happy. This looking for personal validation in another is symptomatic of our incorrect external focus. It is consistent with our deprivation model of being somehow lacking in ourselves, so others must do it for us. It is deficiency love. We must turn our flashlight beam *outside-in* and recognize our own power rather than constantly seeking fulfillment from the outer world. In reality, you and you alone are your own perfect soul mate. No one can complete you but yourself and you will remain incomplete until you do so.

So, we go out into the world and act as if positive relationships were our strong point. By doing this affirming action in the physical plane, we convince ourselves that it's true, and our beliefs can modify accordingly. This is symbolically celebrating the changing of your viewpoint and the starting of "a new day" by affirmative action. It's making a positive and fun change in your life with an act where you reward yourself or otherwise signify a new and upbeat situation. This a part of your *attitudinal healing*.

For example, reflect again on the unpleasant bodily sensations which had signified your former belief system. Focus on their opposites and create consistent action such as taking yourself out to lunch or dinner. You are your own primary relationship and therefore this is a very important date. Get dressed up, treat yourself to a new hairstyle, buy some new clothes in the current fashions, change from somber to bright colors in your wardrobe, buy yourself a present and have it gift wrapped with an appropriate card of affection and inspiration addressed to yourself. There are no limits to this type of process; just let your imagination and ingenuity take over. It really becomes fun; a delightful game in which you are the creator and the recipient. Use this process as a guide for other areas of your life in which attitudinal change is appropriate, such as health, prosperity, family and others. Remember our beliefs and their resultant thought forms are magnetic — they attract to them the experience of reality that is consistent with them. Conversely, sampling the desired physical reality works *outside-in* to align the appropriate underlying beliefs.

4. Sometimes when we have difficulty, not in locating and discreating a negative belief system, but in recreating a substituted positive belief system, we're not clear enough to be able to restate it in belief terminology. We then use the *outside-in* approach by starting on the outside and acting out the experience of reality that we would like to happen. This also is what we would like to see in our mental movie, our new fact pattern. In personal growth and human potential movement terminology it's called pacing, where we act out or copy the desired physical experience. This will help us to discreate the dysfunctional belief, more clearly define the desired new belief, as we imitate the desired physical result and recreate it's underlying belief.

For example in the area of prosperity, let's assume that we, like a number of other spiritually oriented people, have located a belief which says, "High consciousness spiritual people are not supposed to be concerned about monetary rewards." The physical experience of life resulting from this belief system is little or no money, unemployment or a low paying job, a battered automobile, and a life of financial austerity. If someone has carried these beliefs and their resultant experiences for a long time, it is very difficult to get the "feel" of what financial prosperity would be like. It's difficult to convince yourself that it is possible because you have no *experiential* frame of reference.

One way to do this using the pacing approach is to go down to your local bank and purchase a Federal Reserve bag of pennies. A full bag is $50 face value and ½ bags are also available. As you go into the bank, play the role of the watcher and expand your awareness to observe the monied surroundings. Notice the people going to the tellers windows and receiving money. If an armored car is making one of its deliveries, notice the heavy bags of money which are delivered. When you pick up your bag of coins, feel it's weight and imagine yourself as being a part of this monetary environment. When you get home, open the bag of coins and pour them out in the middle of your main living area. Observe how shiny they are, what a large pile they make. Sift through them with your hands and feel their texture, hear the sound that is made while you're doing this and notice the mental pictures and dialogue of "money" that is running in your mind. Now sit down in the middle of the pile of money and begin to scoop up handfuls of it and pour them over your head just like you are swimming in an ocean of prosperity. As you do this shout or sing gleefully, "I'm rich, I've struck it rich, the Universe is drowning me in money." Do this if you can while you are facing a mirror so you can watch yourself smiling and laughing and showering yourself with riches.

Do you get the picture how you can act out any number of scenarios to help you cross the bridge from theory to a hands on experience? The more you do it, the better an actor you will become and the more realistic will be your performances. It will become progressively easier for you to dissolve the encrusted negative belief systems and to substitute your more desired scenarios.

5. There is nothing wrong either with watching TV programs, reading books or attending movies that are positive and inspirational, and have happy endings. These may be criticized by your ego mind or the societal negative conditioning as being unrealistic fairy tales, but I assure you that being exposed to them is beneficial. You know it already because "they make you feel good," which is a sure clue that they are good for you

on many levels. Seek them out in your program guides, book review sections and movie calendars. This is, again, a reprogramming process where you are gradually reconditioning yourself *outside-in* from a negative to a positive outlook. As the momentum gradually builds you will find that it not only becomes easier, but you naturally gravitate to these forms of expression as your experience of life similarly, becomes more affirmative.

And don't worry about missing out on important news events or losing your knowledge of what's going on in the world as you cut back on the amount of your attention to new media. It's all one great big soap opera and if you have watched any of these serials you know that there is just a continuation of basic storylines. Most of the major soap operas on the networks run for years, and if you had gone into a cave in Tibet for a 6 month prolonged meditation, you could return, turn on your TV and resume your understanding of the storyline without much difficulty. Remember it's our immersion in our soap operas and related dramas that create our negative belief systems and activate our self imposed suffering in the first place. We so identify with them that we forget we are not the actor, but merely the watcher of the action.

More positive input is available to us from numerous other sources. Do you read magazines and books? You probably do, and if you were to look at your collection they no doubt would include numerous stories, both fiction and fact that highlight intense personal dramas, violence, fear and anxiety, and related aspects that reinforce the consensus viewpoint that existence is painful and the world in which we live is dangerous. Do you watch movies on TV, through home videos or the old fashioned way in the neighborhood movie theatre? Again be aware of the offerings you have seen lately. This is all potential inputted programming for your thinking mind and will be reflected accordingly through your negative belief systems into your experience of physical reality. There are numerous examples of positive, uplifting, and inspirational magazines, books and movies which are every bit as pertinent and entertaining as their darker counterparts.

I'm not suggesting that one should withdraw from the

everyday world of consensus negativity and thus insulate themselves from any negative input; rather, a balancing is more appropriate until you are able to reach the desired level. Once this occurs, your actual experience, which is attracted by the input level, will confirm the extent to which you are leaning in the negative—positive wire walk. It's like seeking a certain temperature of water and you have an old fashioned fixture with separate faucets for positive and negative. Perhaps the negative has been running full blast for awhile and naturally the scalding hot water is quite painful. When you turn on the positive faucet and the cooler water is mixed in, the end product becomes much more pleasant. When you gain sufficient awareness and skill to not only keep turning up the positive faucet but also close down the negative, the desired temperature is achieved much more rapidly. It's a matter of playing with the faucets of input until the optimum fine tuning is achieved.

6. A great deal of helpful input is also available "out there" in the form of lectures, workshops, and other organized presentations of personal growth and higher consciousness that are available in person, on video tape or on audio cassette. You can personally attend and experience them first hand, or through the magic of the technology of your video recorder or tape player have the lecture—workshop experience "at your fingertips" in the comfort and convenience of your private residence. You can also fine tune the sensory input which you receive when driving in your car everyday by turning off the loud music or news talk shows that are preset in your tuning bar, ejecting the heavy metal rock cassette, and instead using the time to vicariously attend one of these consciousness raising learning experiences through the many audio cassettes that are available.

7. We can elevate our blocked energy by "constructively" converting it into more appropriate usages. This is what Feud described as "Sublimation," where certain negative drives, impluses or other input are expressed in socio-personally acceptable forms.

For example, suppose a person is terrified by voices that are constantly chattering in his head. They tell him that life is dangerous and fill him with fear. Rather than act these voice stories out as crimes or become psychopathically lost in them, he experiences their impact by becoming a writer of stories whose themes are horror, suspense and the occult. Their vividness and impact upon him is converted to a positive usage for creativity and this transformed energy is released back into the Universal flow so that it can cycle creatively once again. In the process, he is consciously dealing with his fears and allowing them a suitable outlet. Edgar Allen Poe and Stephen King are literary giants whose works are these dialogues of their own minds.

In these areas of reality re-creation, since we know to a reasonable certainty that a certain amount of societal negative belief pollution is present in our busy western world lives— it's important to try to have a daily balancing with external positive input, as well as the regular quiet times where we rest inside, within the calm and well-being of our higher selves. As we transform the negative energy into positive, it flows back into the totality of our conscious awareness.

ACCESSING THE INNER YOU

There are many excellent basic reference sources in the form of books, audio cassettes and video tapes that provide detailed information to help us learn how to go inside and get in touch with the wisdom and creativity of the greater part of our being. We can learn to quiet down and relax (still the mind and body), concentrate and contemplate (create meditative/altered states of consciousness), image and visualize (waking dreams), incubate and recall dreams (explore the sleep state), and otherwise go beyond our usual boundaries of knowledge and perceived limits. All of us have some working knowledge of them, both intellectually and experientially. They are of assistance in turning our world *outside-in*.

Going inside requries a cessation of the mental activity that keeps us focused so much on the externals. This quieting process is the essence of the conventional meditative and contemplating disciplines. Paradoxically, the way we quiet down is to "quiet down," in body and mind. We do this by allowing the stopping of our mind-body processes, not forcing them. That is why the most basic meditation is ceasing physical activity, sitting quietly, stopping the input of outside noise, closing the sensory doors (eyes, ears, etc.) and observing the in and out of our own breathing. As thoughts come in, we just observe them as another sensory phenomena, don't get lost in them, and let them go. This is a constant flowing process which suggests to us the true nature of life as the ebb and flow pulsations of energy that it is.

The most direct route inward is by following the breath. This is the basis of Vipassana Meditation, where we just feel the in and out of the breath past the nostrils or the rising and falling of the diaphragm as we inhale and exhale sitting with eyes closed, straight backed and relaxed. When our attention wanders, we just gradually and gently bring it back to the breath. By using "Buddha Breaths," which is deep breathing from the stomach rather than the chest, we are assured of both a rapid relaxation response and the opening of an access channel to our innerness. Deep breaths in through the nose, a pause to hold, and exhalations through the mouth to empty will bring the fastest results. The key is to keep returning to the flow of breathing, no matter what dance the mind does. Eventually, thoughts quiet and you are in that magic inner place.

Instead of revisiting this material, I'd like to suggest other, more unconventional pathways to our inner selves, that can be utilized in addition to the basic source material. This will be adding to your personal bag of tricks so that the stubborn doors of inner perception, that have been shut for so long with years of encrusted corrosion can finally yield. This lubricating process allows them to slowly open and reveal the real world. It's like applying a dose of New Age WD40, as the label says, to "loosen rusted parts, sticky mechanisms, protect, and stop squeaks."

A current technology that opens the gates to the inner self is the use of "mind machines." They are scientific devices that combine sound, light and electrical impulses to induce altered states of consciousness and heightened awareness. Deep meditative states are achieved on demand, just by using these devices. They allow the user to experience the feeling sensations of peak experiences without the tedious preliminaries of yoga postures, long periods of meditative preparation, or assorted austerities.

Perhaps the most popular are the "hemi-synch" cassette tapes developed by Robert Monroe, founder of The Monroe Institute in Faber, Virginia. He worked for years on what neuro-scientists call whole brain thought. Ordinarily the left and right hemispheres of the brain act independently. However, in deep meditative states of alpha and theta consciousness the brain shifts into a single balanced rhythm and its creative power is enhanced. Monroe's theory involved training the brain by feeding into it through stereo headphones certain sounds that duplicated brain wave patterns of peak experience, thus enabling the individual to actually experience that mental state. His terms "neuro-entrainment" and "frequency following response" refer to the binaural beats that the brain hears, causing it to produce its own internal electrical signals that equalize the man-made input and reproduce the alternation of consciousness. These altered states can be repeatedly visited by the individual; the more they are experienced, the easier it is to return.

His new "H plus" technology goes even further, creating an access channel through which his tapes input key function commands (like affirmations) to maximize one's potential in many areas of mental and athletic performance, enhanced concentration, creation of desired realities, sleep restoration, psychotherapy and self-healing. These functions can be activated on demand during everyday life situations.

These are also some personal techniques we can play with to modify the rigid structures of our lives, reveal areas of ego clinging, and open to our inner dimensions. What we do is surrender control. It's only illusory anyway since all form changes. Only the inner realm has a flowing permanency. What we are doing is going behind the structure of our ego.

Record the physical sensations, feelings and emotions, and the voices of your mind in reaction to the following:

1) Alter your usual sleep patterns—stay awake when you would normally go to sleep and observe the slowing of your inner processes, to the point of extreme fatigue. Wake up earlier then usual, at a time when the world around you is asleep and observe it and your reactions to it. Make time to allow your-

self to sleep longer then usual and notice the effects both in your dream activity and your waking reality.

2) Disrupt your daily routines of activity—don't follow your standard schedule; rather, be spontaneous and follow your intuitive urges to action.

3) Practice sensory deprivation—late at night when it is very quiet, listen to the sound of silence; slowly walk around in the dark using the sense of touch rather than eyesight; if necessary cover your ears and eyes to shut out distractions; lay in a tub of warm salty water, in the dark, allowing yourself to just float; or experience float therapy in a commercial tank. It eliminates all sensory input since there is no seeing, hearing, tasting, or smelling. When the water is heated to your exact normal skin temperature there is also no tactile sense of inside or outside, it's just all one beingness.

4) Live in no-time, like the gambling centers of Las Vegas and Atlantic City where there are no clocks on the wall. Remove your wrist watch and any other time devices. Let the concept of time float away as you spend your usual daily activities without consulting it, focusing on your inner timing mechanisms and the expansion of your awareness.

5) Be a fly on the wall—anonymously broaden your physical frame of reference by experiencing new and different activities, with no purpose other than to observe yourself and others and notice your reactions. This could include the bar scene, attending religious services, witnessing a political rally, being a spectator in a court room, going to a 12 step group meeting, being an unknown volunteer sitting in a hospital waiting room, eating a free meal at a rescue mission for the homeless, or any other activities that are not a part of your regular pattern.

6) Find your joy—Ask yourself what it is in life that fills you with exuberant energy, excitation, and feelings of fulfillment when you experience it. What is it that brings goose bumps to your flesh and tears to your eyes? Sit quietly and listen for answers to these questions and then go out into the physical world and try experiencing the suggestions.

THE TRAP OF ENLIGHTENMENT

We have been exploring concepts and procedures to open into the greater part of ourselves, beyond physical form as revealed by our senses, beyond the boundaries established by our ego thinking mind, and beyond the rigid models of consensus reality. As we turn our world *outside-in,* and become still enough to tune in to our quiet inner spaces and explore them—remarkable things begin to happen. We gain a knowing awareness of our true origins and the purpose of our visit to the earth plane. We begin to appreciate the subtleties of the ethereal veils that separate the visible world of physicality and the vast invisible world of spirituality. We see the delicate interplay between both in all of our thoughts, feelings and actions and understand the interconnectedness of all beings throughout this multilayered and multileveled existence. As we become more adept in quieting down and noticing, as our mind stills and our heart opens, we visit other dimensions of existence. We explore altered states of consciousness that bring us in touch with the tremendous powers that we really possess as energy essences of pure conscious awareness. When we begin to acquire these powers, we may however get stuck in quicksand, and detour on our journey of awakening.

The Hindu word *siddhis* means spiritual powers. It relates to the seemingly other worldly abilities of yogis, mystics and spiritually oriented beings. They include the ability to conjure up meaningful visions, explore the past and foretell the future, move physical objects with the powers of their mind, communicate with levels of existence beyond the physical, demonstrate amazing acts of imperviousness to injury, self-healing, extraordinary strength and mobility, and numerous other phenomenon which consensus reality labels as the occult, wizardry, mediumship, and various types of extrasensory perception. These types of powers are available to everyone in varying degrees of intensity, when the enlightenment process of opening occurs.

Simultaneous with the arrival of these powers, which one may experience as new and expanded insights into "the way of things," the Universe presents another one of its numerous ironies. Just as the enlightenment process involves a melting away of the boundaries of the ego self/thinking mind so that we may get behind it and open to the greater realms of our being, the acquisition of some of these powers revitalizes the ego, reconstructing it's boundaries even stronger then they existed before. We begin to view ourselves as very special beings, compared to the others out there who don't meditate as well, have not received the deep insights and awarenesses of the game of life that we have, and who remain lost in their everyday dramas. We become elitists, lost in a new drama called "spiritual materialism." This is a term coined by Chogyam Trungpa Rimpoche. It describes a tendency when we are opening to our spiritual powers, to focus our ego on *better, more,* and *mine* the same way we had previously at the beginning of the game in the material world of shiny cars, higher paying jobs and more fulfilling relationships. If we do succumb to this trap of spiritual materialism we are, for a time, right back where we started in the game of having to become aware of it, extricate ourselves from it by acknowledging our spiritual powers as just another of the phenomenon of life, and then move on in our journey. Paradoxically, the greater the powers that we open to, the more likely we are to become lost in them. Again, the key to finding our way back is heightened conscious awareness.

Another spiritual trap is our tendency to only want to experience the positive aspects of life—avoiding pain or negativity at all cost. But this is just another suffering scenario, where we cling to the happy times and avoid anything less. This is an outgrowth of both our western either—or programming and our impatience to get through with our mortal attachments and achieve enlightenment. Bo Lozoff says,

> "The new 'affirmation' spirituality seems to be based on fear and denial of everything that hurts. But true spirituality allows us to experience ALL of life—not just the easy parts—with fearlessness, respect and honesty."

So it is not un-spiritual to be sad, angry, or depressed; or to be otherwise off center in response to certain events. To do otherwise would be failing to honor our humanity. The key is to allow ourselves to fully express all our emotions when necessary, but at the same time not get lost in them and lose our awareness of the overall lessons of the experience.

GURUS, TEACHERS AND OTHER FRIENDS
ALONG THE WAY

As our personal awakening process occurs and we begin to open more and more to the greater parts of our multi-dimensional existence, we oftentimes become "hooked" on a particular spiritual method or the message of a particular teacher. This of course is no less a mental attachment whose clinging creates suffering than any other of our dramas of life. It is just hidden in the wilderness of spirituality, as if the mere fact that we are pursuing enlightenment means that we cannot get lost in our Journey.

As a general principle we are all aware that everyone we meet in our dance of life is a teacher because all our experiences (which usually involve other people one way or another) are the curriculum for our learning in this school of higher consciousness which we call life. This includes the wisest and kindest beings we have ever met, as well as the insensitive monsters who orchestrated the Holocaust. From this perspective everything is a teaching and everyone is a teacher. However, when we go from generalities to specifics, it is harder for us to see what is happening and we often become totally lost; abdicating our conscious awareness, personal power, and rights of self-responsibility and free choice to a charismatic person who becomes our leader or guide. It happens again and again in our society; in the form of political dictatorships, religious fervor, terrorism, cults, and various "movements" in which people are blindly led around like sheep, listening to the seductive melodies of a particular Pied Piper.

The Universe, in its continuing perfection, usually orchestrates a scenario where the exalted leader ultimately falls from his lofty pedestal and his followers, disillusioned but much wiser for the experience, are given the opportunity to continue on their Journey of Awakening. We see this repetitive pattern throughout history, and in more recent times it has been illustrated by the fiscal excesses of the TV evangelists and the sexual misadventures of the Eastern Gurus.

While one may question whether the media ministries were ever really spiritual or were just elaborate con games trading upon our inherent inclination to seek the comfort and knowledge of answers to the great mysteries of life—the fallen Eastern teachers present a direct case in point for illustrating how easy it is to become lost in the physical world of the senses.

These teachers were raised and trained in the East, where their way of life was one of renunciation and celibacy, so as to heighten their ability to resist the distractions of the physical world and open to the greater truths of being. They came here to share their teaching, recognizing that the lives of Westerners were dramatically unconscious, because most of us lived in our external material world of the physical senses. That was reality as far as we knew it and we were constantly lost within the webs and shadows of our personal dramas. The Eastern teachers had wonderful knowledge to share and the expansion of their teachings has greatly assisted in our individual and collective awakening. However, in the 1960's and 1970's, the era of sexual liberation and independent free thought, many of the teachers succumbed to the seductiveness of the "sex, drugs, and rock-n-roll" revolution. They had never experienced anything of its type in their sheltered, almost monastic lives. Their fall and similar ones of other so-called leaders, serves as a constant reminder to us of the poignancy of our common plight, as we seek to live effectively in our everyday world of sensory overload and at the same time maintain sufficient conscious awareness to not become totally lost within it.

Ram Dass makes an interesting distinction between gurus and teachers. He describes a true Guru as "a cooked goose," a being who is done with learning the lessons of attachment and nonattachment in the physical world. The Guru has graduated from school and, like the Buddhist Bodhissava, remains on the physical plane to assist other beings in gaining the freedom of enlightenment. The existence of the Guru is one of total equanimity. Whether he lives on the street with a begging bowl or has a five room suite in the Hilton is of no consequence—He is a true exemplification in body, mind, and spirit of "neither this nor that." The teacher, on the other hand, may certainly provide valuable insights and convey true wisdom in helping us to see the nature and purpose of our Journey more clearly. But he is still in school, just like the rest of us, although he may be a senior and we are still freshman. He can get lost from time to time in his own dramas of power, manipulation, and the material world of the senses. When this occurs we can gain an important positive lesson of increased compassion for ourselves and all others on the Path of Enlightenment, including our teachers. It illustrates our human frailties in this difficult Journey which are a part of our shared common struggle and interconnectedness.

PART VI
ADDITIONAL PROCESSING TECHNIQUES

We know the truth has been
Told over to the world a thousand times;—
But we have had no ears to listen yet
For more than fragments of it; we have heard
A murmur now and then, an echo here and there.
        —EDWIN ARLINGTON ROBINSON
        (Captain Orsig)

All experience is an arch to build upon . . .
What one knows is, in youth, of little moment;
They know enough who know how to learn.
        —HENRY ADAMS
        (Education of Henry Adams)

No man is free who is not master of himself.
        —EPICTETUS (Discourses)

PART VI
ADDITIONAL PROCESSING TECHNIQUES

OUR BAG OF TRICKS

We've already learned a number of methods that incorporate our theoretical understandings into practical experience. "Processing" is a term coined by the human potential movement of the 1960's and the 1970's to signify the personal work we do on ourselves to turn our world *outside-in,* and focus on our personal psychodynamics of thought and behavior. In the words of Abraham Maslow, the founder of humanistic psychology, we "become self-actualized" by fully using and exploring all of our talents and abilities. According to Maslow, self-actualizers are people who fulfill themselves and positively impact their world, using their potentials to the fullest. The goal of life, when viewed from this perspective, is to maximize one's potential and resources in a manner that allows you to be self-actualized, which is the higher consciousness of the development hierarchy. In this western context it is the equivalent of the eastern concept of enlightenment.

We all have an inherent tendency toward self-actualization. However, before we can achieve this full use of our potentials and abilities, we usually encounter habitual road blocks of patterned experience which limit our development in the form of the negative layering process that we previously discussed, which causes us to lose our true sense of self. Processing techniques help to reestablish the lines of external and internal communications so that reintegration occurs and we are able to heal ourselves in body, mind and spirit through therapeutic unlayering and positive re-layering so that we are whole and complete again.

CURATIVE—The Five Schools of Psychology

Our objective is to train our minds so that we can move away from our knee jerk reactivity to people, places and events and become more responsive. Responsivity provides us the "magic moment of reflection" so that we can more consciously be aware of and properly interpret in a skillful way the outside event that is occurring. We are always using our physical senses to input and evaluate our experiences of life. What we see, hear, think, taste and touch provides a mass of data that may be initially interpreted on an immediate basis by our biocomputer brain, but to a greater extent our perceptions and actions are governed by the filters of our programmed tendencies. Modern psychology seeks to explain these layers of our programming in its various schools of thought that present their own distinct viewpoints of our personality:

1. The psychodynamic school, represented by Freud, Adler and Jung, emphasizes the subconscious mental forces that drive our mind and thus influence our behavior. It focuses on the biological aspects of these unconscious impulses, complexes and conflicts that arise from our past. As such it emphasizes thinking about what's wrong with us, in the rigid structure of a formalized therapy of analysis.

2. The humanistic school, represented by Maslow, Rogers and Perls, is much more flexible and informal in its focus on conscious thoughts, values and self-growth tendencies. It emphasizes feelings in its essentially positive focus on the here and now of our experience of life.

3. The cognitive school, as advanced by Albert Ellis in his rational-emotive therapy, attempts to give answers to problems using a reasoning process that focuses on an examination of our beliefs and expectancies. Once we understand them and how they are a causative link in our life experience, they can be revised if necessary to assist us in achieving a more positive result because we may respond to a stimulus based upon what response we believe is expected of us.

4. The behavioral school, represented by Pavlov and Skinner, and the social learning theories of Dollard-Miller, Bandura

and Mischell emphasize the similarity and predictability of species based upon the conditioning and behavior modification that occurs through direct stimulus and response for observed behavior. Laboratory research and statistical analysis are used to great measure in this school to reflect findings as to differences in behavior of the same subject according to changed stimuli.

5. The trait school, represented by Allport and Cattell emphasizes the differences between people due to inherited natural tendencies and pre-dispositions to certain types of behavior. These so called traits unify the link between a common stimulus and the responses that result in a particular type of behavior. Once the trait is understood the behavior can be predicted.

The interesting overview of these five schools of psychology is that when considered together as a unified synthesis of theory, they are helpful in explaining, analyzing and predicting almost all aspects of human personal experience. They suggest methods to resolve problems of dysfunction that may occur. However, their basic point of view is trying to assist in curing the effects of the reactive tendencies of mind after they have been triggered and are motoring our resultant behavior.

That certainly is helpful but what we are really looking for is the ability to cultivate some degree of detachment so that rather than being trapped in the reactivity of our sensory input and having to then try to cure it—we equip ourselves with *preventive skills* that enable us to consciously respond to the same sensory stimulus.

PREVENTIVE—Cultivating The Teflon Mind

Personality is defined as, "the dynamic character, self or psyche that constitutes and animates the individual person and makes his/her experience of life unique." We want that unique experience to be as positive as possible. To the extent we can assure it by skillful processing techniques, we can enrich our lives. An essential perspective is the view that *all phenomena is neutral*—it is only the labeling of our minds that creates our reactive emotional states. We want to develop skills of *involved detachment,* so we can fully engage life, but not become lost in its dramas. We want to reduce the stickiness of mind that causes it to grab onto passing thought forms. Instead, we want to create a smooth non-stick quality of mind that allows the free flow of mind-body sensations.

A. I am a movie star—

Consider yourself as a movie projector, and display your dramas on the screen of your consciousness. You use many interchangeable lens filters. Project sweetness and light through rose-colored glass, bright red for anger, yellow for fear, gray for somber scenes, bright blue for carefree happiness, and shimmering white light for the uplift of spirit. The point is you are the controller of the type of move that is being shown, whether it is comedy, drama, horror show, or love story. It's all a function of perceptual filters, and none of it is real—it's all your attitudinal projection. You write the script, cast the players, direct the action, and produce it for distribution in your viewing arena. And you, of course, are its star. At the same time, the greater part of you sits in the audience viewing the show in relaxed amusement, understanding that "I am not my movie; rather, I am it's detached watcher." When things get very heavy and you realize you are stuck in one of your dramas with their familiar conflict story lines of health, wealth, love or family — remember it's only your movie and you are a spectator. Take that significant figurative step back from the action to disengage yourself, go out to the lobby and enjoy a bag of freshly made popcorn. It can be fun to go to the movies even when your drama is the main attraction.

B. I am a camera—

One of the reasons for reactivity is that our thinking mind constantly labels everything. It knows nothing of neutral phenomena like a young child that is first learning to express itself. In many ways we need to try to regain the camera image mind of our childhood. We would see certain patterns of light, energy and form and be told that this was a horse. To us it was not big or small, old or young, beautiful or ugly; rather it was just "horse." The camera image of our sensory impression is similarly neutral until we provide labeling judgements to our mental pictures. We understand our mind doesn't know how to view anything except by being an either-

or judge. However, we can cultivate playing the camera game and lessen our tendency for labels.

Walk down a busy street, take a stroll on the beach, or merely observe yourself in daily commerce and notice how your visual picture input is labeled, like sticky yellow memo notes that cling to the surface of the picture. Next visualize tearing off those labels and throwing them away. What's left is just the visual impression *freed of all adjectives* and returned to its basic neutral essence. Become like the child's camera mind, click-click-clicking away and receiving the input of visual images of man, woman, face, and body.

C. I am a sound recorder—

Our mind not only labels visual input but to the same extent is influenced by auditory stimulus. The sounds we hear, specific words, tone of voice, all present us an impression of phenomena with built in critical evaluation. Like the camera that just takes a neutral picture of what is in front of it's lens, the sound recorder duplicates what it hears without any criticism, judgement or evaluation.

If your boss comes to you and tells you your work is wonderful, your reaction is joy. If you are told your work is lousy your reaction is sadness. If however you just interpreted the sounds you are hearing as, "Your work is gobbledegook"— you are neutral because the sound has no labeling-judging emotional significance to you.

During your daily activities, imagine the words and related sound vibrations you hear are in a foreign language you do not understand. There is no way you can be emotionally reactive to them because their content has no cognitive significance to your thinking mind. All you can do is become a better listener, more aware of the detached neutral space between receiving this input and your mental conclusion that it is non-reactive material.

D. I am a bubble machine—

Our most constant companion in life is the continual bubbling spring of thoughts in which we are constantly immersed. Most people experience fear, anxiety and disorientation when their thought processes are slowed or stopped due to involuntarily occurrences of intense focused concentration or the voluntary pursuit of meditative and contemplative techniques. We know however that the more we can slow our stream of thoughts, the greater can be our awareness that we are not them, rather they are our creation. What we create we can of course discreate by practicing the noticing of our thought bubbles, gaining the awareness of them emerging, rising and dissolving in a constant flow, and then developing the mental ability to visualize this thought stream process whenever we choose. In this way our mind clutter can be cleared away on demand by the pop-pop-popping of the bubbles.

Since one of the essentials for any of these processes is breaking through the layers of seriousness that hamper us from skillfully dealing with them—have fun. Go to a variety store and buy a bottle of bubble solution. Like your childhood days go outside and create bubbles with your own breath, use the wind, or just run along the beach holding the bubble maker over your head. Notice the bubbles as they are formed; their shape, size and iridescent color; see how they slowly emerge, gain form and density, and then disappear—all as a part of their own energy. *Your thoughts do the same thing.* You are not the bubble that you are making any more than you are the thought form that emerges from the bubble solution of your own mind. Just sit back and calmly notice your thought bubbles coming and going, rather than believing yourself to be trapped inside them.

THE MAJOR DRAMAS

As you read these words take a moment to view your life as a series of movie themes, and try to identify the major dramas that have thus far occurred. They usually fall into three basic categories; physical or emotional health, financial prosperity, and family-romantic relationships.

1) Health. We spend most of our lives identifying with our physical body. We perceive that it is us—the home of our sensory mechanisms, the place where we ingest food and drink, the seat of our desires and their pleasures or pains, the home of reproduction, and the place where we think our ego mind lives. Until we awaken to our greater probabilities of existence, our sense of self is equated with our body, and we constantly fear physical death.

From a spiritual standpoint, we know that the physical body is the vehicle for this journey of awakening. It is inhabited by the true us—the energy essence conscious awareness which is beyond the limitations of form, and from which our physical form was created to carry us through this lifetime of experiences. The physical body is an immensely complex mechanism; it's exquisite functioning and dynamically balanced interrelationships usually go unnoticed and unappreciated as we take for granted our state of good health. It is only when illness occurs that we take a more active interest in our body and it's place in the mind-body-spirit complex.

All illness, whether physical or emotional, is a state of mind. Our higher self needs to communicate with us to explore a misalignment of our energy that has occurred which throws us out of balance, and into a space of disharmony. The word "dis-ease" specifically means that we are suffering from a lack of ease, our natural symbiotic relationship has been disturbed, and our higher self needs to get our attention so as to help move us back on course in our Journey of Awakening. Sickness is therefore usually an invited condition (ignoring for the moment karmic instances where illness is a part of one's soul blueprint), a way to force oneself to slow down and become aware of the need to change certain aspects of our life that are creating disharmony.

The higher self will usually first bring forward an idea (a thought form) that you should or should not be doing something. If that communication is ignored, your vibrational rate is slowed down so that you receive this message within your emotional body and feel the emotions regarding the situation, telling you that the change is needed. If action is still not taken and you sit around wondering why you are feeling so depressed or "out of sorts," your higher self slows the vibration even further until it manifests within your physical body. There are very few individuals who can ignore a physical communication such as this, whether it be in the form of an illness, an accident or some type of physical disability. The experience of "disease" is the communication from the higher self. Until it is understood and acted upon there cannot be a healing.

Obviously the more heightened your awareness and attentiveness, the less the necessity that these signals from the higher self to put you back on the right course be manifested physically in experiences of illness. It is only as a last resort, when you ignore or resist these communications that physical pain or emotional suffering occurs.

If we are experiencing good health, it is another one of those clues from the Universe that everything is proceeding well, we are on course and all systems are go. If however some form of "disease" is manifested, we need to contact our higher self through meditation or any other lines of communication and quietly ask, "What am I doing or not doing? What am I grasping for, resisting, or clinging to? What is the true communication behind my physical experience?" Once the questions are asked from a quiet space of inner detachment, the answers will be heard.

2) Prosperity. Our western society has a model of hard work, material achievement, and financial success. This is the goal of life which is ingrained in us from an early age as the puritan work ethic that was brought to this country by the earliest settlers. Our external focus on the material rewards of life is pervasive; indeed, it is one of the major causes of our improper focus of awareness that leads to many of our dramas of suffering when this model is not achieved.

From the Universal perspective, we are all meant to be financially prosperous if that actuality is consistent with the physical reality we are attracting to ourselves through our belief systems. Under the Laws of the Universe, it is our birthright to "live the good life" and to have all the material accoutrements that are consistent with it, as long as we don't become lost in the dramas of owning and acquiring, retaining and possessing, and building and achieving. Paradoxically again, once we abandon our attachments to prosperity and our desires for wealth, we reap the benefits. We receive it all, once we have let it all go!

In many instances we reject the bounty of the Universe without even realizing it. This in turn signals that we don't want prosperity and of course the Universe always gives us what we really want. For example, not a day or two goes by, if we are aware, that we don't see money in the street as we go about our daily travels. A penny here, a nickel there, sometimes a quarter; this seemingly insignificant pocket change is actually a clue from the Universe that we can "change" our financial experience. Many people don't bother to accept pennies, throw them away, and reject the opportunity to possess them when they are noticed. Finding a penny or other coin of the realm is a symbolic offering of prosperity to you from the Universe. It often occurs in serendipitous fashion when you are in doubt or despair, having wavered in your faith that the Universe bends in your direction. It is both a reassurance and a reaffirmance of that trust. It may also be a synchronous occurrence such as your friend repaying a forgotten debt to you in the exact amount that you need, but don't have, for your rent payment. The examples are numerous; the underlying principles are the same — in order to experience prosperity you have to believe that you are worthy and physically welcome each opportunity for affirmation.

3) Relationship. Of all our major dramas, relationship is probably the most highly charged and commonly experienced. Whether we are talking about parenting, families, work environments, friendships or romance—all fall in the category of relationship. In fact we spend most of our lives moving in and out of it in one form or another. Volumes have been written about it, so any in-depth treatment is beyond the scope of this book. However, within the context of "changing ourselves for the better" certain aspects of the relationship drama are significant.

Of all our major soap operas, it is easiest to get sucked into how we should or should not behave or think in relation to others. Our culture inundates us with models of acceptable behavior such as, "proper parenting," "being a good son or daughter," "the beneficent master," "the obedient servant," "my true love," and "living happily ever after." It gets to the point in a relationship that we are constantly *shoulding* on ourselves no matter what we do or how we do it because we are always running into these consensus rules of conduct which are in actuality not remotely applicable to our everyday interactions.

If we are guided, not by arbitrary societal rules, but by the sensitivity, kindness and compassion of our open intuitive hearts, we usually make the right moves and experience positive results. As we open to the greater part of ourselves and raise our conscious awareness, we are guided away from reactivity and into the appropriate responses to any situations. When we act from a space of awareness, acceptance and loving kindness—that is exactly what we receive in return, no matter what the particular factual situation.

In the soap opera of romance, everything is intensified. There are more models, more experiences, more intensity, and much resulting agonies and ecstacies. While substance abuse and chemical excesses get all the current headlines, relationship dependency is every bit as prevalent and insidious. In many ways it is even more of a problem because it's symptoms are often unnoticed, consciously or unconsciously concealed, and lifetimes are spent just assuming that these situations are a given fact of life. How much time have we spent in this lifetime thus far trying to find "that significant other" who will complete us and make us whole? Part of our original misconception of ourselves as being limited, flawed beings was the idea that one of the major purposes of our lifetime quest was to connect with that special other person who would somehow bring us back to the state of contentment and perfection that we experienced while in the womb. This lifetime search for society's model of "mommy" hasn't gender distinctions; rather, it is a common desire for the return to a state of unconditional love. Of course we know that this magical place exists in all of us within our own heart. But we first must gain this awareness and learn to access our own heart space so that we can complete ourselves as individuals. No romantic relationship can be totally fulfilling until we have perfected our relationship with ourselves. Paradoxically again we search and search for our soul mate when in fact he/she is really *us*, hidden within the illusory boundaries of our own human hearts.

We are taught in Western society to avoid close contact with others. Most of us don't touch each other or engage in meaningful conversation where we share our feelings. Ironically, we emphasize sexuality—as if it is a substitute for the missing intimacy in our lives. True intimacy grows out of friendship, in a slow but steady progression of trust, sharing, mutual interest and attraction. One of the best descriptions I have heard of this state of being is, "You touch me, even when you don't touch me."

Isn't it ironic that this most complicated, intense and difficult area of societal interaction is the easiest to experience. No mandatory educational or licensing procedures are necessary to enter into a romantic relationship or to have children; unlike the requirements to drive a car, practice law or medicine, or even operate a store selling books or tapes about relationships. As in all other dramas in our curriculum of life—the lessons to be learned, the techniques for learning, and the resultant achievement of desired purpose are revealed by the journey inward.

4) Detective work. When one of our major dramas is playing out on the movie screen of our consciousness we can use the opportunity to uncover the dynamics of the situation. While we are caught up in the drama we are of course not aware of the true situation. We are reacting to a real and serious sequence of events. When we are able to slow down and stop ourselves, if even for just an instant, and notice what is occurring we are able to step away from the movie screen and refocus ourselves from reactivity to responsiveness. Once we realize that we have become lost in the drama, we have the opportunity to gain the awareness to detach from it, if we choose to, and analyze the situation to determine what caused it and why we reacted in the manner in which we did. Consistent with another of the paradoxical Universal laws, we must first be sucked in to the drama (lost), before we can extricate ourself from it (found).

The scripts of our movie dramas may all be different, since each fact pattern has its unique features; however, the underlying forces at work are the same and the manner and method in which we can gain true understanding of them does not vary:

1) We first must notice what is occurring. This requires us to come to a full *stop,* step away from the action, and observe the drama and our involvement in it on the movie screen of our consciousness.

2) We become again the personal detective, and scope out our physical and emotional reactions to the factual situation. We note the mental chatter that is verbalizing our underlying belief systems. We dissect the experience, peeling away its layers to observe the nature of the linkage between the exterior physical experience, our ego involvement through the words of our thinking mind, our feelings, our emotional reactions, and finally in this reverse processing the core belief

which is consistent with these other steps as their trigger device and prime mover.

3) Are we experiencing any suffering? If not then perhaps we enjoy the drama and our role in it. Perhaps it meets our models and satisfies our desires. However, since we are focusing awareness on the situation, we must ask whether or not we are creating negative karma by its continuation. Are its consequences, as a result of our actions or nonactions, creating a situation of suffering for others? If so, we may well choose to walk away from the situation and let it go rather than assume responsibility for its continuing disharmonious energies.

4) Once we decide to extricate ourselves from the drama and end our suffering or the karmic consequences for others we follow our formulas for reality creation/discreation (awareness, intention, understanding and action) and ending suffering (accepting, letting go, and trusting). This processing all takes place in an instant, once we train ourselves through repetition to become aware of our dramas, notice the extent of our involvement in them, and then make conscious choices whether to continue or disengage.

Our dramas are our stuff of life; the daily melodramas which are our curriculum for learning our lessons of personal growth. These dramas are not good nor are they bad; that labeling process is the work of our ego thinking mind. Rather the continuous ebb and flow of our dramas parallels the pulsations of our own conscious awareness as we move along on our Journey. Our dramas do not bring us joy nor do they cause us pain; rather it is our perception of them and the manner in which we react or respond to them that determines the quality of our experience. Our happiness is directly related to our ability to notice the clinging nature of our mind so that when we next identify with one of our dramas and become lost in it, with the resulting emotional consequences—we can quickly and gently free ourselves, re-establish a quality of detachment, as the conscious awareness that sits behind it all impartially observing the action, and bring our lives back into harmony and balance.

KEN KEYES COLLEGE

Ken Keyes has devoted his life to what he calls "the science of happiness." It represents his integration of eastern philosophy and western humanistic psychology. It also blends the scientific method of research and measured observation with the free-form personal focus of the human potential movement. Add to this large doses of common sense and insights from his personal life experiences and those gained through years of offering personal growth workshops, and you have a complete package of loving assistance. I consider him my mentor in bridging the gap between intellectual and experiential knowing. In this general sharing of some of his processing techniques you will get a flavor of how you can help yourselves to get clear of layered patterns of thought and behavior that impede you from living a self-actualized life.

1) Workshops and Support Groups

Group interactions are very helpful in our *outside-in* process of self-healing. There is a bonding of trust and support which builds as the group participants let down their defenses and share their "stuff" which is creating experiences of suffering in their lives. The group dynamic is an accelerated shared energy which facilitates both the investigation, processing and resolution of problems. My personal experiences in attending workshops at Ken Keyes college in Coos Bay, Oregon, demonstrated for me this phenomena. In the beginning I was wary and fearful, reluctant to open up to the group. When I did, rather than being met with criticism or confrontation, I experienced an overwhelming sense of acceptance and a shared predicament with the other participants. This group support and bonding serve to create a safe atmosphere for intimate personal sharing, enabling the release of long repressed fears and hurts. The cathartic experience was in itself healing. Verbalizing one's demons reduces them from lumbering monsters to harmless little gremlins. When it was combined with processing exercises done within the group context, there was an accelerated understanding and the practical re-learning of more appropriate methods of relating to myself and the people, places, and things around me.

2) The Choice Process

When I arrived at workshops I would know where my suffering was; I was in touch with what types of situations pushed my emotional buttons. However, I didn't know how to work through them in a constructive way so that I would both understand their dynamics and be able to deal more skillfully with similar events in the future. Ken suggests that we can pinpoint the cause of our suffering by asking ourselves, "What does my mind want (what does it emotionally demand) when I am physically and emotionally experiencing the suffering?" This identifies the addictive demand we are acting out. He further suggests that we always have positive intentions behind our thoughts, feelings and actions, even though in a situation of suffering we are using them unskillfully. So he suggests we also ask ourselves, "What is it that I want to feel, or see myself as or to hear my mind tell me when this particular factual experience of suffering occurs." This identifies our positive intention behind the addictive drama.

After zeroing in on our reactive emotional demand to a situation and understanding the positive intention that we are unskillfully trying to achieve, we can go further and use what Ken calls the "choice process" to brainstorm some new and more skillful ways to achieve our positive intentions in which we do not create suffering for ourselves. It is very helpful in dealing with non-personal outside events, the conduct of others and related grasping, resisting, clinging experiences of attitudinal suffering we are creating for ourselves.

We can articulate both the addictive demand that is running us, and the beneficial positive intention we have been thus far unskillfully seeking to achieve by using this statement format:

- *"I create the experience of* impatience and frustration *because my programming demands* that traffic move faster on my drive to work.
- My beneficial positive intention is to feel calm and relaxed.
- Three new ways to achieve my positive intention are:
  — Take deep breaths and meditate while waiting for traffic to move.
  — Use the time to listen to a personal growth lecture or new age music on cassette.
  — Find a different route to work."

The choice process assists when our suffering relates to trying to change something in our lives that is not changeable at the moment. By gaining the awareness of new options to achieve the positive intention desired, we can eliminate the pain we were experiencing from beating our heads against a stone wall of our demanded event not happening.

3) The SOS Process

In relationship dynamics, one of the keys to working through the problem areas is the ability to consciously communicate with the other person. This resolves most misunderstandings because they are usually due to improper communication. It further enables the individuals involved to arrive at mutually satisfactory compromises to seemingly unresolvable situations. When one's initial emotional reactivity to a situation has cooled down so that calm discussion is possible, Ken suggest "sharing of space" or "conscious confrontation."

The SOS process eliminates the feeling of separateness from the other person who you originally misperceived was "doing it to you." After you cool down from your reactivity, you gain the awareness that you have been doing it to yourself because of how you choose to emotionally deal with the situation. In a spirit of helpful togetherness you can clear the air by sharing with him/her the fact that you have been creating your own experience of suffering because your mind was demanding that something happen *other than what is*. You further indicate what your positive intention is, then merely listen, without interruption or comment, to the other person's response. You then repeat what they said so that there is no misunderstanding and then thank them for listening to you.

If you and your partner are sufficiently evolved, you can even go further through the process of "conscious confrontation" to not only share what's bothering you but also try in a loving way to effect a change that you desire. You first express how you feel in the particular situation that is causing disagreement. You then communicate your concern about what might or might not happen in the future if the situation continues. You then express exactly what it is you want to happen. Thus far you have done essentially the same thing as the SOS process.

Then, in an effort to understand your partner's point of view, you suggest what you think they want to accomplish and ask them to agree or disagree. Again this is done in a very calm and loving manner so that the confrontation does not create additional suffering. When your partner agrees with your interpretation of what their goal is, you then present some alternatives that you suggest may be acceptable to both of you. In this spirit you can use your ingenuity to find an alternative where both of your objectives can be achieved. In so doing you have defused the situation, ended the current experience of suffering, and created a positive method for future interaction.

4) Unfinished Business

There may be situations where your suffering relates to unfinished business with other people. You may be carrying around anger, guilt and assorted hurts with no apparent outlet because communication with the other person has not occurred. It may be impossible due to their intransigence, physical distance, a situation where it is not appropriate to speak with them, or they may no longer exist on the physical plane. The SOS and conscious confrontation processes can still be done. If you are alone you can visualize the other person and imagine their responses with as much seeing, hearing, and feeling intensity as is necessary. In a group setting, you can partner with someone and they can play the role of the person to whom you are directing the communication.

You can also make your peace with unfinished business by writing a letter in which you express, perhaps for the very first time, everything that you would want to say in person. Many times, the emotions involved in your personal suffering experience are so intense or difficult to isolate that verbal communication may be difficult. In many instances however, the writing process is an effective way to locate the true nature of your suffering, it's true emotional basis in what your mind is demanding, the actual desires which you are trying to fulfill, and realistic ways in which you can either achieve your desires or consciously let them go. This writing process is used to crystalize your thinking and access the true dynamics of a situation prior to actually utilizing verbally processing techniques.

5) Hugs

The Ken Keyes workshops begin and end with hugs, satisfying

and nurturing the open heart needs of self-worth and loving kindness imprinted so deeply within us. Living as we do in a society characterized by alienation from self, solitary living and separateness—the hugs remind us of our inherent oneness and re-kindle our spark of mutual caring and compassion.

Physical contact between beings is the most natural method of communicating love and acceptance, a return to the warmth and security of the maternal womb. Our traditional handshake greetings are really a method to maintain distance between the participants, a throwback to earlier times when people meeting on the road would extend their arms to verify that neither were carrying weapons. Once passing their mutual test, they were allowed to proceed on their way.

I remember returning to Miami after two weeks of hugging and sharing meaningful communication in a workshop setting. I automatically greeted friends with a hug rather than a handshake, and asked them to share their feelings rather than just talking about football or the weather. They had a difficult time dealing with this uncharacteristic social intimacy. It threatened their models of customary interaction. Gradually they became more receptive and they opened to their own heart space that had been closed down by the daily layering of their lives in the fast lane of contemporary life.

It is probably obvious at this point that the number of processing techniques to locate, analyze and uplevel personal suffering are limited only by one's imagination and ingenuity. The more they are explored, others are revealed and the beneficial results increase accordingly. I would be interested in your sharing with me "goodies" from your personal bag of tricks, that have proved informative and helpful to you in your personal Journey to alleviate suffering and experience more lasting happiness. I'm always adding to my bag of tricks, too.

THE NOBLE EIGHT-FOLD PATH IN ACTION

When we were looking at the nature of suffering we talked about Buddha's Four Noble Truths which explain in very pragmatic terms that suffering is caused by our inherently insatiable desires, manifested through the clinging nature of our thinking mind. The cure for suffering was to transcend these cravings and mental attachments by adopting a way of life in thought-word-deed which provided a middle way between a hedonistic life of constantly seeking fulfillment of all desires, and the opposite extreme of asceticism, where one withdraws and renounces all worldly pleasures. Both extremes carry with them inherent suffering, whereas the middle way allows one to live in the world with a degree of detachment, where you can enjoy the moments when your desires are satisfied, but at the same time accept situations in which they are not. You can let go of cravings and clingings of mind except for being aware of them as they rise and pass away in the flow of moment-to-moment conscious awareness.

The noble eight-fold path proscribes a manner of conscious living, the following of which creates an openness of mental awareness and an appreciation of ones personal dramas of life, but also cultivates detachment from identification with and absorption into them. The eight-fold path, both individually and collectively is a positive processing for personal growth and higher consciousness in accordance with the Laws of Universe which are its underlying framework:

1) Right belief—Personal truth is both the objective and guideline for our conduct. We need to continuously examine our belief systems and the link of causation from them through our feelings and emotions and actions to the physical reality that we are experiencing from moment to moment.

2) Right resolve — We must consciously examine the consequences of our actions to be assured that we are not harming others. We should attempt to be calm at all times, moving from reactivity to responsiveness by preserving a magic moment of reflection before we act in thought, word or deed.

3) Right speech—One should be as aware of ones verbal actions as the physical actions. We should attempt not to use harsh language, lie, slander, or otherwise send out disharmonious verbal energy. The old adage that one should "turn your tongue over seven times before speaking" is especially applicable in emotional situations. By doing so, one allows for conscious detachment and reflection.

4) Right behavior—One should not steal, kill, covet, or do to another anything which one may later regret or be ashamed of. This again requires one to "stop, look, and listen" allowing reflective thought before possible disharmonious action.

5) Right occupation—Is the manner in which you earn your livelihood consistent with your truth? If not, conflict is a necessary result and your life and its consequences will be disharmonious. If your occupation provides you with feelings of joy and fulfillment, you are being guided by your truth; if not, the lack of these positive qualities is a clue to you that this inconsistency needs correction.

6) Right effort—One should always do their very best, no matter what their task or occupation, always seeking good and not evil. One must always think about the consequences of one's actions. By doing whatever you do, to the very best of your ability, you are acting impeccably and not generating any negative karma or disharmonious consequences so long as your ultimate objective always considers the effect of your efforts on yourself, others, and your community.

7) Right contemplation—This is the thought and assimilation process of the Four Noble Truths. By gaining both an intellectual understanding and a feeling awareness of them, one is led along the path of acceptance, non-suffering, and conscious living.

8) Right concentration—This is the mindfulness of moment-to-moment awareness which, if followed, leads to inner peace and living a life of truth and happiness. This moment-to-momentness allows one to live in the perpetual here and now, not clinging to the past or grasping for the future. This is a life that welcomes all experience without judging some as good or others as bad. This is a life of "neither this nor that;" rather, it accepts everything as being an integral part of the continuous and ever changing flow of life.

BENEFICIAL MADNESS

The more work we do on ourselves, the clearer is the fine line we can perceive between the labels of insanity or enlightenment. Our breakthroughs often occur at the point where we think we're going crazy. Our "cracking up" is really just breaking out of our old shell of habitual patterns of thought. When the imprisoned energy escapes, we feel pulled and torn because our former foundations of belief, emotion and behavior have shifted and we are exploring new territory as we pierce the veils of conscious awareness.

When asked to try to verbalize what we are feeling, we may say, "I'm going out of my mind." And this is very true. As we leave the realm of the ego and the intellect, pushed by the rush of transitional energy, we are able to expand beyond our previous boundaries into the greater part of our conscious awareness where our most dynamic growth occurs.

POSITIVE NON-ACTION

From time to time we all experience mood swings. We attribute them to many causes, such as pre-menstrual cycles, phases of the moon, illness, and changes in our personal dramas. Usually, since we are such habitually reactive creatures, we judge such times as negative and we struggle to create a sudden change so that we can rapidly return to what we consider to be a more acceptable feeling state of equilibrium.

Our instances of feeling vaguely out of sorts, disoriented and not quite centered are often associated with boredom or general restlessness. Our usual reactivity urges us to resist by running away from this discomfort through "getting out and DOING something." In actuality, this urging to action is our ego self offering us a *distraction* from the introspective awareness necessary to regain our balance. So we resume our external focus and get lost again in another personal soap opera.

These mind-body sensations are really positive clues signalling us that our energies are recycling, seeking a more creative outlet for healthy growth. Our creative juices are simmering, so instead of turning off the burner and leaving the kitchen, we need to stay put and complete the cooking process. Our most positive action is to *do nothing* and watch what is happening.

Cultivating this quality of patience is difficult, but it allows us to notice how our energies are self-adjusting if we allow them to find their appropriate new level. They change of their own accord, according to our own unique internal plan, when we successfully resist the urge "to do something." Rather than viewing our predicament as negative, we can welcome it with eager anticipation because new growth through learning experiences and insights is about to come into our lives.

Positive change is usually preceded by a form of chaos. The water has to boil before the new food of thought and experience is fully cooked and ready to be sampled. It is how we move from scene to scene in our continuing life drama. The energies themselves will dictate to us what is appropriate. Our awareness of them is our best guide to a correct response.

We can even combine a stance of consciously going in and going out, where we use the agitation energy as it builds up, for our renewed momentum in moving outward to help others. We can then create an alchemy process that converts negative to positive energy. This energy flows more freely as we use it in altruistic activities. We and the objects of our assistance are both benefited.

FOLLOW YOUR BLISS

Thoreau said, "The mass of men lead lives of quiet desperation." The reason is they are not doing what their inner selves want to do. Rather, they are following the required roles of their external consensus reality. In short, they are not heeding the advice of mythologist and historian Joseph Campbell who suggested we all should "follow our bliss."

The hardest thing about this advice is recognizing what our bliss is. What is it in life that fills us with eager anticipation, energizes us, brings a smile to our face and a springiness to our step? What is it, when we are doing it, that has no sense of time passing or our separation from it as the doer? The answer will be a description of our bliss—that activity which brings joy to our life.

Most of us live our lives in a time zone marked by weekends, holidays, vacations, or other special events. The in-betweens, which comprise the majority of our waking hours, are the drudgery of our daily routines. This is our "Thank God it's Friday" mentality. Think about it. What sort of lives do we really lead if our daily activity is something from which we are forever planning escape?

Sit down in a quiet place and ask yourself, "If I could do anything in the world I wanted with my time, what would it be?" Then write down the responses you receive. Don't worry about who is talking. Perhaps it's your intuitive self, perhaps it's your ego thinking mind, perhaps it's your layers of programmed tape running off. It doesn't matter here, because the objective is to just compile input for consideration.

Look through the list and see how it compares with your current activities. Assuming there are some things on it you are not currently doing, either on a regular or part-time basis (a safe bet to be sure), pick out any one item and do it as soon as possible. When you do, notice your physical and emotional reactions, and write them down. Think about where that experience fits into your daily life, how it compares, and how you felt while doing it.

Discuss with your family, friends, and acquaintances this exercise. Ask them the questions and notice their responses. You may be surprised how few people have thought about it—they have been following habitual patterns for so long there is a consensus assumption of validity for the status quo.

Repeat this process until you have a strong feeling sense of where your bliss is, and then go for it. Free yourself from that trap. Become aware of your freedom, power to act, and the many alternatives that exist. Create the daily movie of your choice.

PART VII
HOW DO WE CHANGE OUR WORLD FOR THE BETTER?

If you have knowledge, let others light
their candles at it.
>   —*MARGARET FULLER*

In nothing do men more nearly approach
the Gods than in doing good to their fellow men.
>   —*CICERO (Pro Ligario)*

He who bestows his goods upon the poor,
Shall have as much again, and ten times more.
>   —*JOHN BUNYAN (Pilgrim's Progress)*

PART VII
HOW DO WE CHANGE OUR WORLD FOR THE BETTER?

THE POOL OF SHARED CONSCIOUSNESS

On the broadest level of consciousness, the energy from all our thoughts, words and deeds flows into the vast ocean of shared consciousness and from that expanded place influences our world. Everything we individually do or don't do is from this perspective vitally important. It is all an integral part of the collective energy of the Universe. And this pool of consciousness receives our perpetual input, from the beginning of time to the current now moment. Thus the cosmic scales are always in a delicate balance, constantly actualizing in our physical world of form various potentialities of possible experience. This interconnected energy field contains the imprinted traces of the past, a repository of previous experience which is available to each succeeding generation. Each drop of new input impacts it so that it is always expanding and seeking new expression. Thus, the more we raise our own consciousness through positive growth, the higher is the resulting vibrational rate of our world.

This concept has been articulated in psychological, scientific and spiritual terms:

1) Psychology—The Collective Unconscious

Carl Jung's "Collective Unconscious" is a transpersonal realm containing traces of thoughts, feelings and experiences evolved from prior generations. In this synoptic view of Universal consciousness, it is a shared pool of memory which contains all as-

147

pects of past behavior. It is beyond personal experience, but is comprised of it. It is a mass undifferentiated source where all minds merge. While the specific memories of past individuals are not active per se, there is a sort of averaging of the basic forms of thought, idea and image into "archetypes" that tend to mold or shape the contents of our current conscious experience.

Jung's theories, while revolutionary and progressive, did not provide answers to how this common pool of memory was tapped into by present generations. This problem of inheritance was addressed later by the realm of Science through its studies of Quantum Physics and Genetic Biology.

2) Bell's Theorem and Morphogenetic Science Fields

Quantum theory goes beyond the usual viewing and experiencing of the objects of the world as separate phenomena interacting with each other in time and space. Rather, it emphasizes the essential wholeness and connection between all energy, removing all dimensional boundaries. While we can view intellectually so-called distinct phenomena, in actuality they are really just ripples in the same pond.

Physicist John Bell experimentally demonstrated that a pair of particles moving in opposite directions were correlated to each other, even though they had no direct physical interaction. When one changed, the other did also, in a sort of shared non-physical "knowing"—like the coordinated return movement of the ripples returning to their center.

Rupert Sheldrake used his scientific background and knowledge of genetics, embryology and biology to propose his theory of "Morphogenesis" in his book "A New Science of Life." It moves us further toward an understanding of our personal responsibility and affect on the state of Universal consciousness and its reflection in global events.

He suggests that there are spatial structures (fields) which shape and mold the developing embryos of all organisms and, at a higher level, also control their instinctive movements and behavior. These fields have an ordering or patterning effect on organisms, and the structure of these fields is derived from the actual order and pattern in previous organisms of the same species.

"Through the fields, past organisms of the same kind act upon present ones through a process I call morphic resonance. Morphea means form, so morphic means form resonance. Resonance is form upon form. This action occurs through space and time so that things that have happened in the past—past order and pattern—are present potentially everywhere — and present organisms tune into them."

This tuning process, according to Sheldrake, has two elements. The first is genetic DNA which is the coding that enables the organism to create the correct proteins characteristic of its species. They are then organized through the morphogenetic fields to give the organism its form and to determine its patterns of behavior. Under this dual aspect of heredity (influence of the past on an organism), organisms can inherit acquired characteristics from past experiences of other members of their species. The DNA proteins allow the organism to tune into the fields, and then the broadcasted information contained in them is available.

Experimental psychologists proved this theory in rat water maze tests conducted over a period of 35 years, first in America, then in Scotland and finally in Australia. The results showed that the rats got better and better as time went on in escaping from the maze, and this increased ability to learn was transmitted geographically. The rats in Scotland learned quicker than the original subjects in America, and those in Australia learned fastest of all. This increased ability to learn affected all rats of the same breed, whether descended from trained parents or not. They demonstrated this transmission of acquired understanding beyond "usual" concepts of learning.

3) Spirituality—The Hundredth Monkey Theory

Ken Keyes recognizes this massive energy potentiality of experience that is continuously growing and evolving. In Ken's book, "The Hundredth Monkey," he relates the tale of the Japanese monkeys who lived on isolated islands in the Pacific. Scientists provided them sweet potatoes, a taste they liked. The potatoes dropped in the sand and were consumed dirt and all, until one monkey learned to wash them. Gradually, related members of its group did the same, while others continued to consume the sandy potatoes. Then, as all members of the tribe learned this social improvement, colonies of monkeys on other island suddenly did the same—a certain critical number had achieved this new awareness and it was communicated from mind to mind. The added energy of the 100th member of the tribe created the breakthrough that previously had been limited to the first 99 monkeys. At any point in time, one additional drop of input may tilt the scales in favor of a particular form of physical experience.

His "hundredth monkey" theory, which is presented in the context of suggesting how to prevent nuclear war through the elevation of one's consciousness for peace, points out that you may be the next person, the hundredth monkey, who tunes into a new awareness through positive thought and action, providing the necessary additional energy to cause a quantum leap in consciousness that impacts not only your own life but all of those lives that are a part of the entire energy pool. Thus one seemingly insignificant positive thought, word, or deed may make a quantum difference, whereas up to that point the first ninety-nine monkeys of thoughts, words, and deeds had not yet tipped the scale. As Ken points out, "when enough of us are aware of something, all of us become aware of it . . . The appreciation and love we have for ourselves and others creates an expanding energy field that becomes a growing power in the world."

These invisible energy fields of shared experience and alternate potential realities that surround us have historical and present significance. Since they contain the imprinting of all past events that have previously occurred, tapping into them reveals creative genius, like fine tuning a delicate shortwave radio to receive lost transmissions from long ago and far away. Leonardo, Michelangelo and Mozart are examples, as well as contemporary phenomena such as child prodigies and the exceptional learnings of idiot savants.

And from this elevated perspective we see how individual input from our personal Journey of Awakening, as we change ourselves for the better, is inextricably connected with global impact as we are also changing our World for the better. There is a sign at Disney's Epcot Center that expresses this concept: "For all creatures in the web of life, nothing in the Universe exists alone."

EASTERN PATHWAYS FOR A LIFE WELL LIVED

The systems of Eastern Philosophy and religion have recognized for more than 2,500 years the reciprocal nature of the relationships between one's own personal conduct and the society in which one lives. This is clearly reflected in the teachings of Buddhism which originated in India and Confucianism which originated in China.

Buddhism set a standard throughout the East of peace, tolerance, gentleness, love of nature, and a simplistic way of life. It's practical objective of teaching a way to achieve calm detachment from one's human dramas of life and death emphasizes the nature of suffering and the methods to transcend it. It's Noble Eight Fold Path is a way of living which positively impacts the individual and those with whom they live. Although the primary emphasis is on individual detachment from suffering and attainment of personal enlightenment, there also are additional guides that provide an overall system of morality, known as the Buddhist Ten Perfections. They are;

1) Giving—charity and selfless service
2) Duty—mind your own business and do your very best at whatever you do
3) Renunciation—simplify your life and release from worldly distractions
4) Insight—gain conscious awareness, learn, ask questions
5) Courage—be steadfast, trust and don't waiver
6) Patience—accept life as it comes
7) Truth—Stay fixed and determined on your path of awakening
8) Resolution—be secure and firmly committed to your quest for perfect peace
9) Loving kindness—treat friend and foe alike with caring, kindness and compassion
10) Serenity—approach joy and sorrow with equanimity

When one becomes master of himself and attains this "path of perfect peace" by following these precepts, the Buddhist believes that nirvana is achieved and the individual is released from the continuing cycle of physical reincarnations. That individual can then become a bodhisattva who remains on earth to help all other beings to similarly raise their consciousness for awakening and enlightenment.

Unlike Indian philosophy, where the basic idea was to understand the world by investigating the nature of human life and seeking the solution to life's suffering, Chinese philosophy has been more closely connected with personal fulfillment through the cultivation of social and human ethical virtues. There is a double aspect of greatness in Chinese philosophic thought. Inner greatness is a magnitude of spirit reflected in one's peace and contentment and complete experience of life. Outer greatness is manifested in one's ability to live well within the social context of ordinary day to day existence. Thus, rather than being intellectually remote from one's experience of life, Chinese philosophy in actual practice is the same as it's theoretical aspect. The true test of a philosophy from this Chinese standpoint is it's ability to transform it's advocates into better persons. Confucianism emphasizes social humanism, recognizing that a moral life of high ethics is essential for the development and nurturing of one's spirit and the society at large. These social attributes are found by people looking not to the supernatural or to nature, but rather to their own everyday humanity in order to find the principles of goodness and happiness that will guide them. Confucianism teaches a way to live in collective social harmony and obedience through a personal moral code of ethical conduct which focuses on here and now considerations of everyday life. In the *Analects,* the discourses of Confucius are presented in a series of books that emphasize these essential considerations of man's goodness:

1. Human nature is good, and evil is essentially unnatural.
2. Man is free to conduct himself as he will, and he is the master of his choice.
3. Virtue is its own reward.
4. The rule for individual behavior is: what you do not want others to do to you, do not do to them.
5. Man should strive to become a superior man.

Confucius also taught his 5 constant virtues which he equated with a life well-lived:

1. Benevolence, which is always to think first of what is best for all concerned.
2. Righteousness, which is not to do to others what you would not want them to do to you if you were in their place.
3. Propriety, which is always to behave with courtesy and respect towards everyone.
4. Wisdom, which is to be guided by knowledge and understanding.
5. Sincerity, which is to be sincere and truthful in all you do.

HELPING AND SERVING

In our journeys of awakening as we travel our personal paths of
expanded awareness and higher consciousness, we eventually face
the dilemma of when it is appropriate to move from focusing on
personal issues to the larger arena of compassionate action on
community and global issues. We ask ourselves as we turn
outside-in at what point am I sufficiently evolved to take my expanded personal consciousness out into my community through
positive social action?

We know we must first work on ourselves and explore our own
psychodynamics in order to build a firm foundation for growth. We
know we can't have satisfactory relationships with others until we
perfect our relationship with ourselves. But at what point in this
cooking process can we take the cake out of the oven and share it
with our starving fellow beings? When are we ready to go "out
there" and help?

One extreme of "when" suggests that we must first complete our
personal journey of awakening and become totally quiet in our
minds and peaceful in our hearts. We will in that fashion make the
world a better place. By creating peace within ourselves we, on a
collective energy level, are creating it worldwide.

A second extreme suggests that we must act to help right here
and now no matter what our personal stage of evolution, because
there is no time to waste and there is so much to be done.

Ram Dass recognizes this dilemma of compassionate action and
suggests:

> "Why not do both? If you're not doing both, you're being sloppy.
> You should be impeccable. If you act to change the world, but
> change it with anger in your heart, you're perpetuating anger in
> the world. But yet if you're only peaceful in your heart and you don't
> do anything, you're not fulfilling your karmic obligations as an in-
> carnate. You know, you are a member of the species and an ecosys-
> tem. You've got to do both and keep them working on each other . . .
> but you don't have to change the world. You have to change yourself.
> In the course of changing yourself, it changes."

Since our everyday life experiences are our curriculum for awakening, we can see that this process of "change for the better" is mutually inclusive, rather than an either—or situation.

Next we ask ourselves, how do I help and serve—what can I do to help? This is exactly the question I asked myself when I had reached a point, a few years ago, where I had opened spiritually in my personal growth process to a point where I began to yearn to help others. It was as if another of my veils of consciousness had been unlayered and I wanted to do more than just give money to charity. Rather than being an observer of compassionate action I wanted to participate in helping and serving. I learned that the yearning I felt in my heart was a part of the Universe's process of communication, since we are told, when we are ready, the when and how and where of personal service. It is a spontaneous process that calls us when we are truly ready for it to happen.

For me, the call came as I was sitting one day in my office writing checks to various charitable organizations and leafing through that day's mail. I noticed a letter from a friend that enclosed, without further explanation, a copy of an article entitled "Activism in the 80's" which talked about how to overcome inertia in taking the first step to convert concern into action. It talked about the personal confusion of the author in thinking that he had to become a radical activist in order to qualify as a true helper. It went on to suggest modest, low-key activism as a good starting point for becoming a volunteer.

This was exactly my dilemma at that moment. I recognized that this was an act of synchronicity which provided me the clue to get going with personal service. My personal preference was anonymous type helping rather than standing on a soap box giving speeches or engaging in acts of political dissent. While they were fine for those more radical activists who chose them, I preferred the low key approach. So what I did and what I continue to do is feeding the homeless at a local shelter one afternoon, distributing free magazines and paperback books to hospital patients on another afternoon, and saying "Yes" when helping and serving opportunities are presented to me. At the same time I still balance my volunteerism with other daily activities so that it is integrated within the context of my overall curriculum of life.

The "how" of service may be any or all of the following:

1. Green energy—giving money to charitable causes.

2. Hands on activism—which can be the radical social action of Greenpeace, political demonstrations and work disruption, the low-key approaches which I mentioned, or any other specific types of volunteer projects.

3. Daily helping and serving—This is the concept of karma yoga where we have numerous opportunities in our everyday lives to do things such as picking up street litter, helping someone cross the street, taking time to give directions, and numerous other possibilities of informal daily compassionate acts.

4. Living in a socially moral and ethical way such as is suggested by eastern cultures and our Judeo-Christian society, so that as Gandhi said, "My life is my message."

The "where" of compassionate action is only limited by one's own ingenuity. Most communities nowadays focus on volunteerism. There are volunteer service agencies providing a clearing house of numerous types of helping and serving activities. Shelters for the homeless, hospitals, old age and retirement centers, and numerous other facilities offer the opportunity to be of assistance.

Interestingly enough, after we come out of our personal shells and experience the shared warmth of community that helping and serving provides, we gain insight as to the main reasons that people are involved in these activities. It is not so much to be praised as a do-gooder, to become saintly, or to achieve any particular goal; rather it is for the mere doing of the act because it provides an opening of our caring heart. When we do these things we feel so good we intuitively know that we are doing the right thing. Just by the doing, we have opened to the greater part of ourselves and the feelings of joy and fulfillment that we experience are the confirming feedback that we are right on course.

SEVA

Seva (pronounced "say-va") is a sanskrit word meaning service. The sense of volunteerism which is becoming mainstream in America is incorporating the Seva ideals which recognize that simple human compassion is the foundation of service. It is both a means to relieve suffering in our world and at the same time a method for personal spiritual awakening.

As our desire to help and serve grows, it is energized in many new areas of compassionate action. Whereas at one time the "when" or "how" of service was the main issue; now it is the "where." People less frequently think of helping in terms of "should I do it;" rather they want to help but need assistance, direction and structure as to where they can channel this desire so as to become involved in a meaningful volunteer activity.

The Seva foundation was formed in 1978 for such reasons. It's work over the years curing preventable blindness, refugee assistance, reforestation, native American education and other compassionate action programs embodies the growing tide of conscious helping. Seva also promotes and encourages the formation of local community groups to serve and support each other, and at the same time take their compassion and concerns a step further into their own community service projects. The Seva organization supports and encourages the local groups, assists them to cultivate and nurture their individual loving hearts, and directs this heart energy into compassionate action in their local communities.

This represents another evolution of helping and serving, growing from individually fragmented acts of loving compassion to an organizational structuring of social action. This framework assists people who want to serve. It helps them in overcoming their initial western world aversion to being involved with suffering and the problem of inertia, so as to move from thoughtful concern into affirmative action.

What we will no doubt see in the near future are service organizations, probably modeled after the Seva Foundation, that specialize in teaching people to be effective volunteers, so that helpers are mobilized and put to work where they are needed on a daily basis in communities throughout our country. This would provide an additional infrastructure of active placement programs to match the desires and special skills of community volunteers with local needs.

PART VIII
WHAT IS THE NATURE OF DEATH?

Death, be not proud, though some have called thee
Mighty and dreadful, for thou art not so:
For those, whom thou think'st thou dost overthrow,
Die not, poor Death . . .
                    —*JOHN DONNE (Divine Poems)*

Either the soul is immortal and we shall not die,
or it perishes with the flesh, and we shall not know
that we are dead. Live, then, as if you are eternal.
                    —*ANDRE MAUROIS*

Let us not lament too much the passing of our
friends. They are not dead, but simply gone
before us along the road which all must travel.
                    —*ANTIPHANES (Fragment)*

PART VIII
WHAT IS THE NATURE OF DEATH?

THE ULTIMATE DRAMA

This is not the last chapter of the book according to the table of contents, but it was the final one to be written. Like most of my fellow human beings, I shared some resistance, both conscious and unconscious, to investigating the subject of death. It is our ultimate drama—triggering more of our mental attachments then any other of the personal movies that we produce, direct and star in during our physical life time. When we look at our formula for non-suffering from Part V we can see that it's very difficult for most of us to ACCEPT death as it comes, LET GO of the desire for physical immortality, and TRUST that it will all work out for the best. We resist death through avoidance and denial, cling desperately to life and grasp at its recreational and material pleasures, and find it difficult to have the faith that will transcend our fear of death.

If we consider that the purpose of life is to grow by experiencing the openness, acceptance and compassion called "unconditional love" and to learn how to extricate ourselves from identification with our personal dramas—we realize the biggest obstacle is fear. Fear to open ourselves and risk being vulnerable, fear to interact with others and let go of the rigid protection of our ego boundaries, and fear to surrender to the constant flow of change in our lives. Fear is closing down, while love is an opening process. All the suf-

fering that we self-create through the grasping, resisting, and
clinging attachments of our mind are motivated by fear. And the
biggest attachment of them all of course is to this physical body
which is the vehicle through which we travel in our journey of life.
We long for answers that will soothe our anxieties and provide
some calming assurance of the nature of death.

One of my legal specialties over the years has been estate planning and probate administration. The first step in this process is
usually conferring with clients concerning their will, the docu-
ment where they establish how they will dispose of their property
and the beneficiaries who will inherit from them.

Jacob was a long-time client, now 90 years old, who wanted to
revise his will again because of recent marital problems in his
family. He had made changes from time to time and was usually
pleasant, humorous and quick witted. However, his mental processes had now begun to slow and he was often confused, especially in trying to comprehend the complexities of the newly revised estate tax laws. Our telephone conversation seemed endless.
I listened to the changes Jacob wanted to make in his testamen-
tary gifting, explained to him the tax effects, confirmed what he
wanted, and then had to repeat the entire process as he lost his
train of thought and temporarily forgot what had just been discussed. This went on and on until I finally had the necessary information to proceed. As we ended our conversation, Jacob apparently realized the difficulties in communication that had taken
place, paused and said, "Forgive me Martin. I'm easily confused
these days. I don't mean to take up so much of your time." And
then he added with a chuckle, "But you see, I've never died before!"

We all share Jacob's predicament at least as it relates to this current lifetime. Death is the great unknown, the darkest corner of
our warehouse of fears, and our primary soap opera. In levels of
high drama, emotion and nail-biting intensity, it is our all time
award winning movie. On a scale of 1 to 10, we could rate it a strong
15. But even with its immensity in our mental picture of life's "big-
gies," it is still, when viewed with that all important deep breath
of detachment, just another of our dramas.

Why does the concept of death spin us so tenaciously into its
whirlpool of fear and avoidance?

BIOLOGICAL ENCODING

The survival instincts of our biological organs are intense, and for good reasons. There is a basic genetic encoding in the DNA structure of our cells which includes the basic drives and survival instincts for air, water, food, shelter, sleep and sexual reproduction. This is our linkage to the other species in the animal kingdom who share these primal physiological needs. Charles Darwin's basic theories of evolution recognized the concept of "survival of the fittest," whereby the strength of organisms and the ability to adapt to their surroundings was a direct determinate of their ability to continue in the life cycle.

Abraham Maslow, the founder of Humanistic Psychology, recognized the importance of these encoded aspects in his Hierarchy of Needs. He reasoned that human motivation was based upon an ascending level of needs, each having to be satisfied before the next succeeding one could be considered. The first and most basic were the physiological needs which are our biological heritage. This survival imprinting is a part of the construct of the physical body. It forms the basis for our initial total identification with it.

I AM NOT MY BODY

If we cling to the apparent reality of our physical form, grasp for its perpetual existence, and resist the inexorable physical process of decay, deterioration and death, we will suffer greatly. To the extent we identify with our physical body as being who or what we are, the prospect of its destruction through accident, illness or the aging process is terrifying. Fear has a field day with this mental model. If however we consider the possibility that "I" am not my body, but rather the energy essence conscious awareness that inhabits it, like a space suit, during the visit of this lifetime to the earth plane, our perspective is entirely different. If at the moment of death "I" just drop my body, and am reborn into the non-physical realms of higher vibrational consciousness from which "I" originally descended at the beginning of this lifetime, my movement on to the next lessons of a new incarnation after the mandatory period of evaluating the lifetime just completed and assimilating its growth seems a natural progression in my Journey of Awakening. If the real "I" is perpetual, indestructible, all knowing spirit, our fear of extinction becomes ludicrous because only our organic costume dies, while we who were it's occupants continue on our Path. Emmanuel, a non-physical entity channeled by Pat Rodegast assures us that there is nothing to worry about: "Death is perfectly safe. It's like taking off a tight shoe."

In our view of life on the earth plane as a continuum of growth opportunities, the ultimate learning experience is the death of our physical body. As it undergoes the process of life, decay, death, and transformation that is common to all forms of organic matter we are put directly in touch with all of our mental attachments to physicality. It's almost impossible to avoid them since the physical density and vibrations of the three dimensional earth plane are centered in the physicality of form which itself is expressed through our physical senses. We delight in them, and remain externally sense focused through most of our lives, our experience of them being equated with the satisfaction of our earthly desires and our ultimate happiness.

But the greater part of our being, our wise inner self, knows that the energy essence conscious awareness that inhabits and animates our physical form during this lifetime does not age or die, any more than it is ever born at a given point in time. It just *is*—a neutral phenomena—that is not subject to the laws of time, space, or form. It exists outside of our mental constructs of them, beyond the dramas of our world of constant change. Our thinking mind is itself an aspect of form, so it naturally can not view itself except through the same filters of limitation, impermanence and ultimate extinction.

We certainly should honor our body as the home of our true self during our physical lifetime. Just as our eyes are known as "the windows to the soul" our physical body is "the temple of the soul" and we should revere it as such. This means providing it with as much loving care and attention, rest, relaxation, nutrition, exercise and respect as would be appropriate for the vehicle which enables our journey of awakening to be as long and eventful as we choose it to be. The physical body is our sole means of transportation throughout our lifetime and its importance as an essential aspect of the body—mind—spirit triad is absolute. It is our most trusted and faithful companion as we play the game of life, but it is not the true "I." It is our means of transportation through life, not life itself.

THE FLIP SIDE

One of my favorite movies is "One-Eyed Jacks" starring Marlon Brando. In that movie he and Karl Malden are bank robbers in Mexico who are being hotly pursued by the Mexican police. They decide that Brando will stay to fight them off to give Malden time to round up some additional members of their gang to help him to escape. Malden rides away with his saddle bags full of the bank money. Instead of finding help he leaves Brando to be captured by the police while he resettles in a nearby Mexican town and becomes a wealthy landowner and it's well respected sheriff. Brando serves 7 years at hard labor in a Mexican prison, escapes, and then seeks out Malden for his ultimate revenge. When Brando, the bank robber and escaped convict confronts Malden—the country squire-sheriff, he says, "We're just a bunch of one-eyed jacks, but I've seen your other side."

Life and death are flip sides of the same record that is playing on our turntable. If we emphasize one to the exclusion of the other, our life's journey will be out of balance. A life well lived is one where we are not clinging to the past or resisting the future; rather it is one where our life is actively lived in the perpetual now moment. Living in the here and now automatically quiets the mental chatter that creates much of our daily discord, because when we are in our mind we are automatically pasting or futuring about people, places or things. The space of here and now takes us "out of our mind" as a meditation that opens us to the greater part of ourselves that is beyond time and space and the limitations of form.

A wise sage once said that life should be lived as if death is on our left shoulder. Our work on earth may be over at anytime, and when this occurs our soul essence will choose to exit the physical body. To the extent that we have lived a life of denial, fear, postponement of business, or letting ourselves just idle in neutral—we have missed the growth opportunities that have been presented to us. On the other hand, fully engaging life at every opportunity, attempting to be aware of it's multi-leveled nature, playing the game with joy, passion and exuberance, and at the same time being prepared to exit when the time comes with no unfinished business is the epitome of the idea of "holding on tightly and letting go lightly."

THE ASHRAM OF AGING

We have talked before about the circular nature of our journey of awakening. Nowhere is this full cycle more evident then in the observation of the process whereby the physical body is born, grows, matures, decays, and ultimately dies. As we approach what society calls "old age," whatever the number of years that may signify, we notice certain common physical characteristics. There is a gradual slowing down of body-mind processes. Hearing and sight are diminished. Energy levels fall and the amount of physical activity decreases. The skeletal structure changes as the body seems to become smaller. Skin loses its elasticity and becomes more wrinkled. Verbal and written communication becomes more difficult. Overall there is a turning inward, like a *compulsory meditation* where the same objectives of a contemplative life such as quieting the mind, quieting the body and opening the heart are achieved.

As the aging process reaches its ultimate conclusion one notices a completion of the circle by a return to childhood. When the newborn infant came into the world without the ability to see or hear or speak, physically fragile, wrinkled, physically uncoordinated and non-functional—survival depended upon the care of the parents for this helpless organism. The old age phenomena brings the same characteristics of loss of physical and mental function and general helplessness where the children of the elderly person now are presented with the growth opportunity known as "taking care of mother or father." The perfect interconnections of the circle of life become evident, as this role reversal takes place.

This reverse maturation which occurs as one's aging proceeds toward the end of physical life serves to gradually soften and melt away the encrusted ego boundaries that have been built up throughout the lifetime. It's as if there are thick walls that have been constructed and in order for the energy essence conscious awareness to properly exit, these enclosures must be removed. Sometimes the demolition is sudden, such as in the case of an accident, massive heart attack, stroke, or dying in one's sleep. One moment there is a living functioning physical organism and seemingly the very next moment physical death occurs and one's spirit essence is thrust out to continue it's journey in other planes of existence. When this occurs, one's work on earth has obviously been completed and the ego structure does not require the additional learning experiences involved in a gradual wearing away of its confining boundaries.

DISSOLVING THE ENCRUSTED EGO

Often the ego enclosure has to be removed brick by brick, beam by beam, until the necessary open space is achieved. This is commonly seen in individuals who eventually die from a lengthy, debilitating illness. They may at first be hard frozen into their habitual patterns of thought and behavior, but as they are exposed to the fires of pain and worry, deterioration of their physical and mental faculties, and the growing realization of their helplessness and ultimate predicament they have the opportunity to undergo a form of metamorphosis where the rigid ego melts. They may become much softer, gentler and loving beings who are able to open themselves to the doorway to the greater part of their being at the time their consciousness leaves their physical body. My father's death was an example of this softening process.

He was strongly conditioned by his family upbringing and the times in which he lived. His parents came to America from Russia in the early 1900's, escaping religious persecution. They and the other thousands of European immigrants who arrived in the United States at that time had lost all their material possessions and had been deprived of many of their personal freedoms. They brought with them a sense of required personal strength, longing for peace and acceptance, and the determination to translate hard work and dedicated effort into material success. The lifestyle of these immigrants and my father's parents was characterized by personal sacrifice, the expectation of hardship and suffering, an underlying insecurity, and yet a sort of grudging acceptance of it all with good humor and strong religious faith. Their personal dynamics included little communication of inner feelings or outward displays of affection. The extended family, where parents, grandparents, aunts and uncles, and siblings all lived under the same roof was the structure through which their personal dramas of life were lived.

My father, like his father, was strong-willed, often stubborn in his desire to achieve and reach the goals which he set for himself and his family, uncomfortable with meaningful communication and displays of affection, and ill at ease in numerous social situations; a testimonial to his childhood spent working in the family clothing factory rather than learning the social niceties. At the same time he had a wonderful sense of humor, was a sweet and gentle person inside, and went out of his way on a daily basis to help and serve others. One of his personal goals was doing good deeds for others as often as possible which usually took the form of little gifts and bakery goods which he would give to friends, acquaintances and business contacts without there being any special occasion for giving. And he certainly had a strong ego structure which, to a certain extent, enabled him to achieve and succeed in the face of much adversity.

Most of my father's life was free of any serious illness. He was a very active person who viewed every day as both a challenge and an opportunity. In his busying pace of "doing things" he tended to push himself to the limit. He suffered a massive heart attack in December of 1987, rallied and was released from the hospital, and was re-hospitalized and released on a number of subsequent occasions as his strength gradually weakened. He died at home in June of 1988 at age 80, one week after what was his and mother's 53rd wedding anniversary.

His death process epitomized the ego softening process which finally culminated in his spirit being released. I visited with him almost daily during that 6 month period of his illness, and I was acutely aware of his struggles to control his life, resistance to the physical and mental processes that were occurring, and his periodic grasping to hold onto the habitual patterns of thought and

behavior that made up the familiar layers of his personal programming. Gradually, despite his resistance and periodic
thrashing about in opposition to his predicament, he became
softer and quieter and mellower so that near the end he would
hardly talk. Instead he would uncharacteristically allow me to
stroke his hair, hold his hand and tell him I loved him.

THE PROCESS OF EXITING

Much research has been done into the concept of death, including it's physical characteristics and the common states of mind
that appear and disappear as recognizable stages in the process.
The more one quiets down and achieves a degree of detached
awareness, the more these states can be seen as the most interesting of all observed mind-body-spirit phenomena.

1) THE MEDICAL VIEW — Taber's cyclopedic medical dictionary defines death as:

"Permanent cessation of all vital functions. The following defini-
tions of death have also been considered;

1. Total irreversible cessation of cerebral function, spontaneous
   function of the respiratory system, and spontaneous function of
   the circulatory system.

2. The final and irreversible cessation of perceptible heart beat and
   respiration."

The medical model of death is of course directed toward the
physical body and its vital functions. Little consideration if any is
given to the possibility that our conscious awareness is a separate
and continuing aspect of existence that is not subject to the time
and space limitations of the world of form. It is not surprising that
we spend most of our lives totally identified with our physical
body. Its every function is verified to us on a continual basis
through the perceptions of our sensory mechanisms and our
thinking mind, all of which are aspects of the same world of form.
All our daily interactions with people provide continuing verifica-

tion that their identity is also that of a specific, separate, and identifiable physical form.

The medical community recognizes that the physical body is a temporary vehicle, subject to the ravages of illness, accident, and the natural deterioration that occurs through time. In this respect it is able to be of great assistance in its ability to make necessary repairs, provide replacement parts, improve the quality of the functioning of the motor apparatus, and to generally keep the vehicle running in good condition so that it can continue in it's journey down the path of life. However, since the physical body is equated with our identity, each time there is a malfunction, we experience the psychological trauma inherent in the possible termination of our existence. This annihilation is both unthinkable and unspeakable, and therefore we tend to become totally consumed in the medical soap operas of life.

Death also represents the ultimate frustration for doctors. Their whole orientation and avowed purpose is to save lives. If they do, they are deemed successful and if the patient dies there is a sense of failure. But since everyone ultimately experiences the death of their bodily form—no one gets out of this game alive—the medical community must constantly face not being able to meet its modeled expectation of winning. This has serious repercussions for the psyche's of the doctors as well as their patients.

2) THE EASTERN VIEW—The Tibetans are great observers of the process of death. They have studied for centuries the detailed transition periods or "Bardo's" that we sentient beings undergo in our endless chain of births and deaths as we learn our lessons of physical existence.

1. The natural Bardo of this life—our conscious awareness temporarily inhabits our physical form.
2. The painful Bardo of death—our conscious awareness separates from the body.
3. The luminous Bardo of the after death/death state—our conscious awareness merges back into its pure state of "isness."
4. The karmic Bardo of rebirth—our conscious awareness becomes re-embodied corresponding to the karmic configurations which form the basis for the new incarnation.

This is a constantly flowing process of transition which, when viewed from the Eastern perspective, happens over and over again in the mere twinkling of an eye, representing thousands of physical births and deaths. The eastern viewpoint regards all the physicality of life and death as a sequencing of neutral processes; mind-body phenomena that become more distinct and recognizable as the intense light of moment to moment choiceless awareness and detachment shines upon them.

Tibetan monks have specific meditative practices which focus upon death. They sit for days observing with detachment the transition of their fellow monks from physical life. They see the decay and deterioration of the lifeless physical forms which are returned to their organic roots and the feeding of carrion upon these rotting carcasses which have been emptied of their life giving energy essence conscious awareness. To our western eyes this would appear to be the most gruesome of experiences; to the Tibetans however it is the ultimate spiritual learning because it completes the final aspect of being unattached to any aspect of physical form—even if it be ones own body. This dis-identification with form represents the highest freedom of consciousness and the ultimate objective of the human cycle of births and deaths.

The Tibetans recognize that the physical body of a being is first formed by means of the appearance of the five physical elements— earth, water, fire, wind and space. The body is maintained through their continued interaction and it ultimately perishes when they withdraw. As the moment of physical death occurs, they observe how the out-breaths gradually become longer than the in-breaths; water leaves the body, the skin dries and thirst occurs; heat leaves the body and the individual becomes cold; and as the density of physicality leaves one experiences great lightness and all sensations gradually dissolve into nothingness. Finally all thought states cease into luminosity that becomes concentrated in the heart center, as the emerging consciousness completes it's exit from physical form.

Eastern cultures honor the death process as the highlight of one's life, the most meaningful learning experience that one encounters on the earth plane. They are taught at an early age that even death is a neutral event. It is not good or bad; rather it just *is*. The physical body inhabited by spirit essence is organic form, it is constantly aging, and some day must be left behind to disintegrate into dust and disappear back into that from which it came.

3) THE WESTERN VIEW—Death always wears a black hat in our society. We are taught almost from birth that it is our arch enemy and we will be doing battle with it throughout most of our lives. Our great fear of death is due in large part to the consensus denial that exists around us. It is pervasive in our youth-worshipping, goal-oriented society. We don't talk about it, we shield our children from it, and we spend most of our time trying to prevent it. It is viewed as a pervasive negative, especially in a society whose premiums are on physical attractiveness, the pursuit of sensory pleasure, materialistic object success, and the prolongation of physical life. We do everything we can to hold back the aging process and forestall the physical changes that must inevitably occur.

Death is regarded as the ultimate failure. We remove the sick, elderly and dying from the familiar mainstream of their lives by institutionalizing them in old-age and similar medical facilities. Our funeral practices after death seek to negate it as a natural part of the cycle of life. As we worship the vitality of the young and banish the old to oblivion, we continually relayer ourselves with the stickiest of attachments, clinging to physical life and the fear of its end.

OPENING THE WESTERN WINDOW

Elizabeth Kubler-Ross has probably contributed more than any other individual to taking the subject of death out of its dark closet in our western society and exposing it to the sunlight of awareness. A psychiatrist, she began interviewing dying patients in Colorado in 1965, and through her work and its incorporation in her "Death and Dying Seminars" her books on the subject, and her international reputation as an authority and counselor on death, she has provided an opportunity for us to remove our constricting layers of death programming and re-think the entire subject from a more enlightened perspective.

In her early work she described the stages that human beings typically go through in their experience of dying. These stages—

denial, anger, bargaining, depression and ultimate *acceptance* are also present in any intense loss scenario such as the death of a relationship, an occupation, a familial tie or any other previously fixed and seemingly permanent situation. We carry with us very strong layers of programming that fear any change. Since the purpose of our life is growth through change, and since life is a constant flow of changes, our level of anxiety is necessarily bound to be great. Abandoning our old habitual patterns of thought and behavior is like dying, but living an existence without any change is not living at all because there is no growth. From this perspective the dying of the old and the birth of the new in a constantly interconnected process is an absolute precondition for life.

Ironically, even though hospitals in our western society provide the best medical care and are the customary places in which people die, this very fact contributes to the continuing fear and denial viewpoints. Since for most people the most frightening thing about dying is the feeling of being alone and having to face the great unknown without most of the familiar aspects of their everyday existence that sustain them in times of turmoil—dying within a hospital can contribute to the sense of losing one's total identity and existence. On the other hand dying in familiar home surroundings, a hospice facility, or an alternative terminal care facility—surrounded by loving family, friends and other growing beings, can eliminate most of the fear involved in the death process and transform it into a profound and positive learning experience.

Elizabeth presents this viewpoint of how to end one's life as a part of the journey of awakening to higher consciousness in her book "Death—The Final Stage of Growth."

"Death is as much a part of human existence, of human growth and development, as being born. It is one of the few things in life we can count on, that we can be assured will occur. Death is not an enemy to be conquered or a prison to be escaped. It is an integral part of our life that gives meaning to human existence. It sets a limit on our time in this life urging us on to do something productive with that time as long as it is ours to use. . . . if we look at that from a different perspective, then we can see that it is the promise of death and the experience of dying, more than any other force in life, that can move a human being to grow. All of us, even those who have chosen a life of non-growth—of playing out the roles proscribed by others—feel within our inner most selves that we are meant for something more in this life then simply eating, sleeping, watching television, and going to work five days a week. That something else, that many can't define, is growth—becoming all that is truly you and at the same time, more fully human."

SURVIVAL OF CONSCIOUSNESS

Do you remember Jacob's predicament, shared at the beginning of this chapter? Since he had "never died before" he had no frame of reference to attempt to answer probably the most persistent question we human beings ask during our lifetime as to whether or not life continues after the death of our physical body. What if death is not the end, but rather a continuation of the life-death process, viewed not from a polarity extreme of either this or that but rather as one integrated and continuous cycle of existence? There is clear, compelling, and persuasive evidence that answers this question in the affirmative.

HISTORICAL EVIDENCE

Philosophers throughout the ages have pondered the mysteries of death and sought a solution, upon the theory that one can never fully understand the significance of life until one clarifies the meaning of death. All the major cultures and ideologies of the world throughout recorded time have suggested the possibility that death is not the end of existence; rather consciousness continues in a different form after the physical body is no longer able to function. Acceptance of these beliefs in a continuum of life has had to be made primarily through faith, although there have been a number of saintly beings in every age who have claimed to have a direct experience of universal life and love which has convinced them that "conscious awareness" is not restricted to living within a physical body. It has been called by many other names, including psyche, self, mind and being. It is most often equated in spiritual terms with the soul of man, which passes on into a multi-dimensional realm of existence after physical death. In the words of Goethe:

> "I am convinced that the soul is indestructible and that its activity will continue through eternity. It is like the sun, which to our eyes

seems to set at night, but has in reality only gone to diffuse its light elsewhere."

There are numerous other examples of historical opinions that our existence is perpetual, of which these are some:

"Surely God would not have created such a being as man . . . to exist only for a day! No, no, man was made for immortality."

Abraham Lincoln

"Death, be not proud! A short sleep, and I will wake again, eternally. A short voyage, and I will meet my Maker face to face . . ."

John Donne

"That day which you fear as being the end of all things is the birthday of your eternity."

Seneca

"There is no death! What seems so is a transition."

Henry Wadsworth Longfellow

"When I go down to the grave I can say, like many others, 'I have finished my day's work!' But I cannot say, 'I have finished my life.' My day's work will begin again the next morning. The tomb is not a blind alley; it is a thoroughfare. It closes on the twilight, it opens on the dawn."

Victor Hugo

There is also some common sense affirmation. We know our individual consciousness is not a part of our body. Its existence continues after parts of the body are removed by injury or by disease. It exists independent of what we think of as "us" in our illusion of reality. As a practical matter there is no actual proof that our consciousness is extinguished by death. There is in fact tangible and credible evidence at our disposal that it continues.

EVIDENCE FROM THIS SIDE—NEAR DEATH EXPERIENCES

In recent times a phenomenon has been reported by thousands of ordinary people with no particular religious background or knowledge of metaphysical phenomenon. It has resulted from new areas of consciousness research that relate to advances in modern resuscitation techniques which have "brought back" many people who have nearly died as a result of injury or illness or, even more dramatically, were pronounced clinically dead and survived. Their experiences have common and recurring aspects and have been well and widely documented by respected authorities in the field which began with Elizabeth Kubler-Ross, were expanded by Raymond A. Moody, Jr. in America and Kenneth Ring in Britain, publicized by the Continuum Foundation, and in general have become widely accepted by researchers in the medical and behavioral sciences. In fact a new scientific discipline called "Circumthananatology" has been created to cover these studies of near-death experience, and a worldwide organization called "The International Association of Near-Death Studies" was formed to record the phenomenon, headquartered in Philadelphia. It provides persuasive evidence of the indestructibility of consciousness, answering the human quest through the ages for some type of reliable prediction, that was suggested in the words of Omar Khayyam:

> "Strange, is it not, that of the myriads, who
> Before us passed the door of darkness through
> Not one returns to tell us of the road
> Which to discover we must travel too."

There are numerous common aspects of the near death experience that have been reported by the researchers who have interviewed people with otherwise totally different physical, cultural, emotional and spiritual characteristics. The common profile of how the near-death experience tends to unfold is set forth as follows from the work of Raymond A. Moody, Jr., the primary force behind the emerging awareness of this phenomenon.

> "A man is dying and, as he reaches the point of greatest physical distress, he hears himself pronounced dead by his doctor. He begins to hear an uncomfortable noise, a loud ringing or buzzing, and at the same time feels himself moving very rapidly through a long dark tunnel. After this, he suddenly finds himself outside of his own physical body, but still in the immediate physical environment, and he sees his own body from a distance, as though he is a spectator. He watches the resuscitation attempt from his unusual vantage point and is in a state of emotional upheaval.

After a while, he collects himself and becomes more accustomed to his odd condition. He notices that he still has a "body," but one of a very different nature and with very different powers from the physical body he has left behind. Soon other things begin to happen. Others come to meet and help him. He glimpses the spirits of relatives and friends who have already died, and a loving, warm spirit of a kind he has never encountered before—a being of light—appears before him. This being asks him a question, non-verbally, to make him evaluate his life and helps him along by showing him a panoramic, instantaneous playback of the major events of his life. At some point he finds himself approaching some sort of barrier or border, apparently representing the limit between earthly life and the next life. Yet, he finds that he must go back to the earth, that the time for his death has not yet come. At this point he resists, for by now he is taken up with his experiences in the afterlife and does not want to return. He is overwhelmed by intense feelings of joy, love, and peace. Despite his attitude, though, he somehow reunites with his physical body and lives.

Later he tries to tell others, but he has trouble doing so. In the first place, he can find no human words adequate to describe these unearthly episodes. He also finds that others scoff, so he stops telling other people. Still, the experience affects his life profoundly, especially his views about death and its relationship to life.

Kenneth Ring researched the Moody material and then collected additional material of his own which classified the 11 common near death experience characteristics found by Moody into a basic NDE core experience that had five stages as follows:

1. Peace and the sense of well being.
2. Body separation where there is a sense of detachment from the physical form.
3. A transitional stage of entering the darkness.
4. Seeing and entering the light of the "other side."
5. A decisional crossroads where one had to decide to either return to physical life or continue on away from it.

Both Moody and Ring recognize the significance of the replay process. The person is asked if they are satisfied with what they have done with their time on Earth for "good." There is a deep understanding that "good" means the extent to which they have loved. The love referred to is one of understanding, compassion and selflessness.

This aspect of love is seen again and again in the revelations of those who have had near-death experiences. A pervasive theme is the gaining of a realization that the special purpose of our life on Earth is "learning to love" ourselves, others and mankind. This brings us to our true state of grace and growth.

NON-SCIENTIFIC EVIDENCE FROM THE "OTHER" SIDE— CHANNELING

The growth of scientific evidence that human consciousness survives physical death has been paralleled and intensified by the New Age concept of channeling. It is a communication process whereby pychically receptive individuals are able to channel information from non-physical entities through themselves, in much the same way that we would tune in a radio or television receiver to the frequency in which a broadcast was being continuously transmitted. It is in fact suggested that we are all channels, and have the ability to tune in to the wisdom, teachings and insights of the disembodied spirit energies that occupy non-physical dimensions of awareness.

Today there are literally hundreds of these Spirit Teachers who are communicating to us on the physical plane through their various communicating channels in an effort to help us more fully understand the nature, significance and purpose of our lives. One of their primary subjects of focus is of course death, since they recognize that we have great difficulty in extricating ourselves from the grip of its conditioning on us and our society.

What follows is an excerpt on the subject of death from the 1987 book "Aids: From Fear to Hope" which is a compilation of channeled teachings offering insight and inspiration concerning life in general and the phenomena of Aids in particular. Its spirit teachers provide the benefit of their wisdom like a panel of experts presenting a personal symposium of knowledge:

Understanding The Process

Kyros

Many people on your planet fear the exiting process which you term "death." In fact, many humans, once they become aware that this process exists, begin to fear. Usually, this begins in childhood when a pet or relative dies. Adults try to comfort the child by saying, "Tabby or Fido went to heaven," or "Grandma Pearl went to live with God." But because adults oftentimes mourn so and beat themselves up with guilt by saying such things as, "Oh, I wish I had been kinder to Grandma Pearl," or "I wish I had never kicked Fido or left Tabby out in the rain," children receive this as a statement of finality or that maybe heaven or going to live with God is not really such a positive thing.

Very small children, because they have recently entered from another dimension and still have some dim recollection of it, have less trouble with the process of death. As a child becomes more programmed by the outer adult world, this recollection becomes locked within his subconscious mind and he begins to perceive as the adults in his world do.

The problem of fear exists because, as children, most adults were programmed by other adults. They see death as something inevitable, something mysterious, something unknown, something filled with pain, and oftentimes as something final. And many go through their lives with this hidden fear inside them. You cannot fully live and experience your physical journey if you fear death.

Many people who fear death do so because they do not understand at a deep level that they are Spirit, not form. Your form (or body) is only a vehicle to carry you through your third dimensional journey. It is like a spacesuit or life-support system for physical life. If you went to the moon, you'd have to wear a protective suit in order to do your work on the moon. When you returned to Earth or entered a simulated Earth atmosphere, you'd take it off because you'd no longer have need of it. So it is with birth and death. You put on a form when you enter at birth and remove it when you leave at death. It's very important to understand that your vehicle is not you. You are energy which cannot die or be destroyed, but only transformed into a different kind of energy. You are Spirit and Spirit contains individual mind with all its levels of consciousness. This is eternal. You were created out of the God essence and God does not die!

Birth and death in the physical realm are basically the same process. At birth you enter from a higher dimension, and at death you enter into a higher dimension. You think of death as an exiting, but the only exiting which occurs is when the Spirit exits the physical vehicle. Both birth and death are processes of entering.

Many do not fear the actual process of death as much as they fear how they will die, or when they will die, or the physical pain which sometimes occurs prior to the transcendent experience. Oftentimes, how you will die has been determined by your Spirit prior to entering for your own learning, or to teach others or for some karmic balancing. There are no accidents in spite of what you may believe. At higher levels everything has a purpose and a reason and is designed for growth and unfoldment. What an entity does has to do with his mission. No one ever leaves his vehicle until his individual mission is completed.

I do not wish to sound insensitive about the fears human entities have concerning the process of death. I do understand your confusions and concerns. On the higher dimensions, we do not use the term death. We see what you term as death as merely a transition from one dimension to another, from one level of awareness to another. You chose to enter your dimension to learn and to teach so that you might grow and unfold. When this is completed, you move on. Your great Teacher, the Christ, told you about many mansions and showed you that death is an illusion through His resurrection. If only you would believe.

As for the pain and suffering that sometimes precede the moment of transcending, these are the illusions your mass consciousness have created. You have created pain and disease and brought them into manifestation, and it has become part of the mass belief system which continues with each generation. You chose to create a world of duality for your growth and unfoldment.

All negative illusions which you perceive are based in fear. The Master Christ was always saying, "Fear not!" Why? Because He knew fear would bind and imprison you and prevent you from experiencing a full and abundant life. He also knew that fear of anything is ego-based, and the ego's main function (as you know) is to maintain and protect the physical vehicle. The ego knows that once the spirit is released from the vehicle, it no longer has either purpose or power.

I would (if I were able) convince you that there is nothing to fear in the process of transcending. You will be released from fear when you come to the true awareness that the real you is energy and Spirit, not form. You will be released from fear when you reach the awareness that there is no such thing as death in terms of finality. Even form does not die, but merely transforms itself into something new. Ashes and dust return to the Earth Mother and give birth to new life. Nothing dies. All is but transformed. You will be released from fear when you acknowledge who you really are, a Spirit of God, the God Essence. The physical walk is but a short journey in terms of time. There is so much beauty ahead. Fear is released when you align for your own highest good and growth. Fear is released when you replace it with love which is the healing and transforming energy of the Cosmos. This is what your whole purpose is on the Full Circle Journey: to love, to learn it, to teach it, to be it.

I would also like to add that sometimes humans fear death because of their attachments to the physical. It is human not to want to leave those people and things which you love and have found joy and pleasure in. Fear will leave if you will remember that those people you've loved and journeyed with are also spiritual beings traveling in physical vehicles and that when you move to another dimension, your Spirits are even more connected and attuned to one another. You don't really leave them. As for attachments to physical things, they will no longer have meaning in a higher dimension. Physical illusions belong to the physical realm. Part of the learning on the physical plane should be to move toward release from attachments to illusions, to the alignment to spiritual reality. Once you have made this transition, there should be no fear. Joy does not cease at the point of transcending, but increases beyond human comprehension.

So, if you fear death, I would say to begin working on your release from fear. Think of yourself as a caterpillar in a co-

coon preparing to take flight as a beautiful butterfly into a world of sunshine, or as a rosebud preparing to open into bloom. Though you may not (because of current belief systems) think of it as such, it is a beautiful and joyful experience and not one to fear.

Soli

The Higher Self has planned out experiences before you came here. You have agreed to have karma with certain individuals that you were going to incarnate with. You have a plan of action, as it were, a schedule mapped out. But when you come here within the physical body, you find that you have this subconscious mind and its beliefs. You have total free will and choice that allows you to choose to do anything at all. And so, you find yourself moving away from the path that the Higher Self wanted, the path of highest evolution, of greatest growth, of experience-the true path you wanted to have.

If you are too far off course and there is no further way in which the Higher Self can get you back on course, the Higher Self will say, "All right, there is absolutely no point in continuing this lifetime, there is no point in continuing this journey. We're never going to get to San Francisco this way. We're in New York; how can we get to San Francisco? We'll terminate this journey right here." And, so, you decide to leave the Earth plane and you decide that you'll come back and have another lifetime and try again.

Since you are infinite and immortal, you have an infinity of time to do whatever you want to do. If you don't achieve what you set out to do in this lifetime, you will do it in the next life, or in the next, or however long it takes you to do it.

There is no death, my friends. The subconscious mind loves to hang on to the thought that this is all that there is, that once the body goes, it's all over. And even if their lives were terribly miserable, people cling to life desperately until the very last minutes. They cling to it because of their fear of the unknown, their fear that death is far worse than the that which is known.

Death is a matter of rejoicing! How ridiculous are your somber funerals. You should be partying, having fun, and sending the entity on his way with joy instead of with sorrow. When you leave this plane, you are returning to your Spirit brothers and sisters in great delight. They are, in a sense, partying with you. It is fear and the feeling of loss that you will never see that person again, that creates the somber atmosphere at funerals. It is what you are feeling, it is what has happened to you that creates the sorrow, not what has happened to the person who died. You are really crying for yourselves.

Party, my friends, and rejoice that another has returned home for a time, for a bit of rest and relaxation, before coming back into the game again.

When you leave this Earth plane, God does not stand outside of you looking at the life you have led and then punish you by sending you back. Karma does not work that way. Absolutely not! You, yourself, choose your own lives, one after the other. You judge your own selves, and the lives that you have led and you decide to return and balance the energies in some other way. But no one forces you to, my friends. Everything you do upon the Earth plane is free will and choice totally.

Making Preparation

Dr. Peebles

To prepare yourselves for the transition, be in the light. Live in the light. You are all familiar, hopefully, with the term "hospice." The hospice concept is a rekindling of a concept of creating an environment that acknowledges when our physical life is almost complete. It is a place where you attract light, and where all around you live in light, acknowledging the pain, acknowledging the anxieties, the fears, but calling upon the Guides, calling upon the Higher Self.

If you cannot go to a hospice to die, you should continue living at home and being at your job, telling everyone, "We know I am about to make the transition. I am frightened." Or, you must say, "I'm not frightened, and if you are, go for a ride for the next ten years because I will not have anything except joy and light around me, no matter how miserable and hard this is."

Once you make the transition, if you have not been studying soul communication, or if your intellect has been over stimulated, you will possibly go into a state of sleep. If your religious beliefs are such that you would go to sleep until Gabriel blows his horn, then when you leave the body your soul (or Spirit) will sleep. Oh, it will wake up and look around and say, "Oh, I don't hear a trumpet; I'd better go back to sleep," and it will. But eventually it will get terribly bored waiting for that dumb trumpet and then it will wake up.

And it is your choice. Do you want to sleep when you go to the Spirit side? Then develop the intellect. If you want to go pure and straight through, study after your soul, study after the real you. Give up your old ways, give up your prejudices and your fears, give up your anxiety, and allow everyone else the same privilege. When you leave your body behind, you'll be instantly awake, alert and aware, with no period of adjustment necessary. So, you want to be ready when you pass from this body? Start loving yourself, love everyone and everything and be free.

When Illness Is Terminal

Soli

It sometimes happens that when an individual has a strong communication or illness, such as a terminal illness, it can be turned around. Why? Because the Higher Self perceives that the individual is learning something new, has made changes within his other life, has found a way to affect acceptance of that change, acceptance of an understanding of the communication, and is making great changes within his life. He has, therefore, the possibility of a whole new line of experience, as it were, within this dimension.

Then the Higher Self says, "Well, fine. We will have a lot more different experiences now then those which we intended to have when we came here. We do not need to leave now. We do not need to leave and come back within another lifetime. The communication has been understood. The illness no longer needs to be there." And you have what might appear to be a miraculous cure. What has happened is the individual has taken the responsibility for his or her own disease, has done something about it, turned it around and made the changes necessary in their life and has gone forward, no longer with the need for communication.

Sometimes the Higher self specifically chooses a lifetime that will be cut short by a certain illness, for the individual has a karmic need to experience, perhaps, a particular physical imbalance. They may have caused such a physical imbalance for another person in a previous lifetime, or judge those who had that physical imbalance. You may have an individual in one lifetime who constantly judged and made fun of someone with a particular physical disability, and because of that judgement has decided to have a lifetime where he experiences that particular disability. And so he has chosen a lifetime where he experiences exactly the particular projection of energy. And once he has understood that and worked with it, transcended it, then the Higher Self will decide, "Well, perhaps there is no point in staying here any longer. We can leave now. We've had the experience of that particular kind of disability. We will leave and choose a different lifetime."

Again, you will not stop that from happening. You will not stop the Higher Self deciding to end that life. If you find a physical way of prolonging life, and it can be done through mechanical means, then the Higher Self will create what is known as an "accident," which is simply another way of leaving the Earth plane.

Everything, everything is free will and choice. You are never a victim. You are not a victim of disease, you are not a victim of a healer. No one can come along and heal you against your will, against your need. All healing is self-healing. Understand that and you understand everything there is to be known about healing. It is one of the simplest subjects to deal with upon the Earth plane once you understand it, once you understand that you are God. God is not somebody outside of you, sitting in judgement, casting evil spells upon you because you have sinned, because you have gone against the unwritten law. Absolute rubbish, no such thing! You are God. You are God experiencing the Earth plane, experiencing your own creation each and every minute of the day. Once you understand you are God, that you are everything, that you are already one with all life and with each other, that you are always expressing the God force, the God energy through yourself, the I AM experiencing its own creation, then you have no need for illness or disease.

Master Adalfo

People with critical disease often are better than others at living in the now. It comes with an awareness that they don't have an unlimited length of time here. Of course, this is true for everyone. When they begin to face that and acknowledge it, then each day becomes more important. Each day that is left is more important and they become more able to enjoy the pleasures of that day. That is the ideal. Some people fall into that attitude as a natural bi-product of the disease. For others it is more difficult.

For those heavily into the syndrome of lack, they will only see the terrible things to come and not acknowledge the abundance of the moment. It is important to be doing things with these people, things that will help them to see the abundance. Bring them flowers, bring them toys, play games with them, so that for even a few moments, they acknowledge the abundance of that moment. If you do not spend time with that moment, how do you know if you have a diamond or a rhinestone? How do you know what it is if you don't take a moment to look at that moment with your jeweler's loupe? Each one of you could use a jeweler's loupe to truly see the day, all the facets, all the light, all the beauty. In this way, you will also get to see if something is a dud. And, if that is true for you, you could just throw it out and concentrate only on the beauty.

Understanding this often is the gift that you find in children who have any disease in which they are dying. Children get that message faster. They know what they're here for. And they're often here to just teach others to enjoy each moment, to savor the time that they have together, instead of thinking of the time they will be apart.

The Role of Children

Soli

Any child who dies before the age of thirteen is a Spirit who has completed all necessary lifetimes upon the Earth plane. These children chose to return to the Earth plane out of service to the parents who needed the experience of losing a child. Before the age of thirteen, a child is not fully formed and is still a part of the parents' aura and vibration. After the age of thirteen, all the teaching has been done. The child is fully formed and becomes an adult.

As a Spirit, you incarnate within the physical body anywhere between the point of conception and two months after the actual physical birth of the body. If no individual Spirit wants that experience with those parents or within that society, and if no Spirit chooses that body, there is nothing to keep it alive and it will expire after the two-month period, approximately. This is known as "infant death syndrome" in your society.

The same with abortions. In many cases, the Spirit chooses to be aborted. For example, if a woman aborted a fetus in a previous lifetime and felt tremendous guilt about it, then she would choose to have the experience (in another lifetime) of being aborted herself, just to balance the energy, just to make sure that she did not create any difficulty by doing that in a previous lifetime. You see, it is the belief again. It is because of the guilt behind it. It is not the fact that she aborted the fetus that requires her to be punished. She chose to be aborted in this lifetime to balance that self-judgement.

This applies to every action upon the Earth plane, my friends. There is nothing wrong with any act. It is the belief that accompanies it that creates the difficulties. We say again to you, there are no problems in life. There are only events. The problem lies within your mind and how you view the event.

Suicide

Soli

Many who commit suicide do so out of the mistaken belief that death is the end of everything. For it is prevalent within your society to believe that this is your one and only lifetime within physical form, and that when you die that is the end of it; then you have lost consciousness forever as it were. That is a fond hope of many.

It is most important to realize that there is no judgement against suicide from the spiritual perspective. There is no God judging it to be morally wrong. It is not the worst crime imaginable. It is simply one of the experiences it is possible to have. And it is only the self-judgement, the anger, the resentment, the emotions that are carried through that create difficulty for the individual who has committed suicide. Those that are conscious when they take that action are more able to meet with their guides and teachers, more able to be guided into the higher dimension, more easily able to understand who and what they are, where they are, and so they can move on quite readily. It is not the act of committing suicide per se, that creates any difficulty for individuals. It is the state of consciousness and understanding as that takes place.

But, of course, this applies to what we might call a regular death anyway, or so-called regular death not by one's own hand specifically. It still depends very much on the consciousness of the individuals as to how it is experienced. There are many who die a regular death, who believe that that is the end too, and are very surprised to find themselves waking up. There are many who die in consciousness and have a very easy transition. So, we would say, whether an individual chooses to take his or her own life or not is not that important. What is important is the state of consciousness, the understanding.

If the thoughts or ideas of reincarnation are within the subconscious mind, this usually is strong enough to open such individuals to the possibility that they may be in the astral dimensions. Once that thought comes forward, they can then look for their guides and teachers who will assist them from that point on.

On the question of suicide, you must understand that all, in a certain sense, are committing suicide. When you watch someone ingesting substances within their life, it is a little easier to see how they are, quite consciously in a certain sense, committing suicide. For the ingestion of substances, drugs, etc., any addiction is a slow form of suicide. It is a way of not wanting to have to deal with the experiences of life that you have come here to deal with. It is a way of escape and is perfectly reasonable and acceptable. It is a part of the experiences to be had on the Earth dimension. Suicide is only a slightly more obvious form of escape, more catastrophic in the sense of happening more quickly than the slower forms.

Illness itself is a form of suicide. A terminal illness is a form of suicide, for ultimately you are always responsible for your own health and your own life. And if your own health and your own life deteriorates, that is your responsibility, nobody else's. It is impossible to be a victim, so in a real sense, all life ends in suicide. If you see it in that way, then suicide can be seen to be much less a drastic act than you otherwise might see it, not to be morally condemned, not to be judged.

It is for the individuals left behind to bless and release the individual. Do everything in your power, once the individual has left, to speak to them within the spirit dimension, have them begin to understand, even once they have left the physical body, that there is white light around them, that there are guides and teachers. Tell them that they do not have to stay around the Earth. Now, if they have been troubled on making their transition, they will be around close to you, they will hear you. And you can do a great deal in assisting them within that process to move on to their next experience.

If we were to generalize, we would say that yes if you take your own life prematurely, then there will be a need to return in another lifetime to experience those things that were not experienced. But where there is already a terminal illness, and a lifetime is approaching its end anyway, that is much less so, much less of a judgement on self in that case. And there is certainly no judgment from any other entity anywhere in the Universe. There is no spiritual judgment on this. It is most important to recognize this. There is no mortality associated with this. God is not sitting there judging individuals who take their own lives. It is always self-judgement. It is the thought form, it is the belief that creates the experience for the individual within the astral dimension, and it is those thought forms that will keep them locked within that dimension until they wake up to the fact of where they are and what they are doing and what they have created for themselves.

If you are counseling, or working with an individual who speaks to you about their desire to end their life, there is only one answer, and that is the answer to everything. It is unconditional love. It is not for the counselor to judge the individual and their choices. If that individual chooses and wants to commit suicide, it is not the counselor's job to dissuade or persuade. It is the counselor's job to love totally and unconditionally, without trying to impose their own belief systems and understandings on the individual being counseled. We are talking about a saint here because it is extremely difficult to counsel anyone without getting your own personal feelings and belief systems involved. It is almost impossible. But ideally, this is the way it would be. To be with that individual with such unconditional love that whatever they do is perfectly all right with you.

My friends, remember the full meaning of unconditional love. Unconditional love says, "I do not care who you are, what you are, where you are, or what you are doing. What you are doing in your life is absolutely what you need to do and I am here to support you totally in whatever decision you make. I am not here to do it for you, or to persuade you or to agree with you necessarily, but I am here to support you in the decisions that you make." If an individual has made his or her mind up that it is time to leave this dimension (from the ego point of view), you are not going to stop them anyway. If you try to stop them, you are going to create karma for yourself. You may see them end up in greater suffering. What is that going to do to you? Do you want to persuade somebody to not take their life and have greater suffering?

You see again, it is looking at it from this point of view of moral judgement of the society. Somehow there is a great stigma attached to suicide and therefore anybody who allows it or accepts it in another individual is somehow aiding and abetting a crime. What nonsense to believe that taking one's own life must be a crime within your society. You would much rather lock someone up under lock and key and observation 24-hours a day than let them take their own life. Is that a better alternative? It is a very strange attitude within your society associated with death. It stems from the need to hang onto life to the last instant, believing that this is the one and only life that you have ever had and you are not going to have any others, and that therefore you must hang onto it and you must force everybody else to hang on to theirs. So you find medical practice trying to prolong life against the will of the individual, against the desire of the individual instead of allowing the natural course of events.

You might look at it the other way, that if the illness were allowed to take its natural course, death would have been much sooner anyway. It is the prolongation of life through drugs and artificial means that creates more problems within an individual. And so, you might see that if that attempt at healing from the external, mechanical, point of view had not taken place, then the life would have been terminated already anyway. And so, of course, within the consciousness of an individual, there is the dichotomy, "Should I make use of every medical advancement, should I try to prolong my life, or should I allow natural events to take place and leave when body has the need to no longer function?"

In terms of self-healing, of allowing self to be healed, there has to be an improvement in the quality of life, otherwise there is no point in staying on to experience more of the same. If the individual's life is not improving, what is the point of prolonging that life? Do you want to keep somebody alive and in pain and suffering? It all comes back to the fact that the counselor must exercise non-judgement: The Loving Law of Allowance for all things and all individuals to be in their own time and space, giving unconditional love and not judging the individual for their choices.

One who hasn't been there can never really understand what it is like except perhaps by going back into previous lives and remembering some experiences they have had of similar nature. And then, of course, once you do that, then you have much greater ability to accept everyone else around you unconditionally anyway, for then you begin to understand what life is really all about. We would reiterate, as we have done so many times, that there is no other purpose to life but experiencing it. And experience is what you are here to do. If you want to terminate that experience of current ego, that is no great problem. The Higher Self will create another ego personality in another time and space if it feels the need to continue that experience.

A LAST WORD

If in fact we voluntarily leave physical existence when our spiritual blueprint of this lifetime has been completed and we have finished our work—our primary focus should be on living a conscious life while our physical body continues to be inhabited by our energy essence. This means shedding the layers of the onion that have provided the tears of our existence and opening to the inner sweetness of our being.

A friend of mine died recently from a cerebral hemorrhage. After the stroke had occurred he was maintained on life support while his doctors sought to find ways to revive him. When there clearly was no medical hope, the decision was made to withdraw artificial maintenance, in accordance with his previously stated wishes. I was at his bedside when the moment of medical death occurred. As his Rabbi chanted the final prayers, I had the strongest sense of release. It seemed my friend was there in the room, watching this drama through the spaciousness of his conscious awareness, and laughing at how easy and painless the death process had been. Even though he, like most of us, dreaded a death involving lingering illness and prolonged suffering, he had prepared himself through spiritual practices and opened to the greater possibilities of being. I could almost hear him saying, "I'm really free at least."

Elizabeth Kubler-Ross's book, "Death — The Final Stage of Growth" concludes with the following statement:

"Death, in this context, may be viewed as the curtain between the existence that we are conscious of and one that is hidden from us until we raise that curtain. Whether we open it symbolically in order to understand the finiteness of the existence we know, thus learning to live each day the best we can, or whether we open it in actuality when we end that physical existence is not the issue. What is important is to realize that whether we understand fully why we are here or what will happen when we die, it is our purpose as human beings to grow—to look within ourselves to find and build upon that source of peace and understanding and strength which is our inner selves and to reach out to others with love, acceptance, patient guidance, and hope for what we all may become together."

PART IX
WHAT IS THE PROGNOSIS FOR THE 90's?

They shall beat their swords into ploughshares,
and their spears into pruning—hooks; nation
shall not lift up sword against nation, neither
shall they learn war any more.
                    —*OLD TESTAMENT (Isaiah)*

Glory to God in the highest, and on Earth
peace, good will toward men.
                    —*NEW TESTAMENT (Luke)*

We know what we are, but know not what we
may be.
                    —*SHAKESPEARE (Hamlet)*

PART IX
WHAT IS THE PROGNOSIS FOR THE 90's?

AT THE CROSSROAD

As we head into the 1990's, we find ourselves in a tremendous vortex of accelerating energy that highlights significant stages of ending and beginning. This sense of acceleration is tangible. We can feel it in our daily lives like observing the sand descend from the hemispheres of an hourglass; as it nears completion it appears to move faster, with a kind of reckless, yet deliberate urgency.

We have completed a yearly cycle, entered a new decade, are approaching the new millennium of the year 2000, the 26,000 year cycle of the equinoxes is ending, and our sun and solar system are completing a gigantic rotation around the central sun at the center of our galaxy which is described in the Hindu Veda scriptures as "the cycle of the yugas."

THE NEW AGE

This sense of accelerating energy can be felt by all of us as we are buffeted to and fro with the winds of change and our learning experiences and opportunities for spiritual growth are accelerated. We are being exposed to ever higher frequencies of cosmic energy. They seem to be tearing us away from the familiar moorings of our habitual patterns of thought, belief and action, as they seek to free us from self-created prisons of limited consciousness and thrust-

ing us into the unknowns of the future. We feel restless and unset-
tled, mindful of sweeping change that seems to accentuate our
everyday lives, somewhat wary and anxious, but at the same time
filled with tremendous energy and nervous excitement about the
future. It's as if we are pioneers who have arrived at a frontier
called "The New Age."

There is a tide of cautious optimism as it becomes more mainstream. By no means yet consensus reality, it nevertheless is accelerating at a compounding rate. As it gains momentum, a popular song echoes this sense of quickening, "Well it's all coming
together, I'm on some kind of roll, Baby it's all coming together, I
feel it down deep in my soul."

This "New Age" is a societal opening to the greater possibilities
of identity and existence; a coming "home" to the underlying wisdom of the ages that has been within us but was overlooked as we
became lost in our personal dramas. It has been defined by pub-
lisher Jeremy Tarcher as:

> "a metaphor for a process of striving for personal growth through
> which millions of people are trying to become more fully awake to
> their inherent capacities.
>
> . . . At the heart of New Age thought is the idea that humans have
> many levels of consciousness and that, with the exception of a lim-
> ited number of spiritual geniuses throughout history, we essen-
> tially live in a walking sleep that keeps us from a balanced, harmo-
> nious, and direct relationship with God (however you understand
> that concept), nature, each other, and ourselves.
>
> Broadly stated, this world view is;
> (1) the everyday world in our personal consciousness is the mani-
>   festation of a larger, divine reality;
> (2) humans have a suppressed, or hidden, higher self that reflects,
>   or is connected to the divine element of the Universe;
> (3) this higher self can be awakened and take a central part in the
>   everyday life of the individual;
> (4) this awakening is the purpose or goal of human life."

These concepts, rather than being new date back to the earliest
expressions of man's inherent spirituality and have been called
"the perennial philosophy." Whether or not we as individuals and

members of the earth community take them to heart may determine to a large extent whether or not the human species will survive.

The same technology that freed man from the constrictions of a rural way of life has evolved into the prison of our contemporary urbanity. As we have entered the threshold of outer space and the possibility of other inhabited worlds, we have, in our careless war against nature (assuming limitless resources and creating massive toxic wastes), fouled the footpaths of our home planet.

EDEN OR ARMAGEDDON

> We have in the past been forced into reluctant
> change by weather, calamity and plague. Now
> the pressure comes from our biologic success
> as a species. We have overcome all enemies
> but ourselves.
> —JOHN STEINBECK

The odds being quoted on whether or not Mother Earth and it's inhabitants will survive the end of this century are even money—50%/50 %. The doomsayers point to the following scenario:

A) The Earth (Gaia) is dying, due to man's failure to honor it as a living organism. The earth has become increasingly toxic due to overpopulation; pollution of air, water and food sources; and man's interference with the natural order of nature. The factors contributing to our terrestrial genocide include the following:

1. The (deforestation) stripping away of the great rain forests.
2. The pouring of industrial waste into the air and the seas.
3. Increasing air pollution from auto emissions, industrial smog and smoke.
4. Ineffective disposal of nuclear waste and leakage of toxic chemicals.
5. Acid rain.
6. Increased levels of atmospheric radiation caused by holes in the protective ozone layer, created by the release of fluorocarbons and related chemical wastes into the atmosphere.
7. Greater extremes of worldwide temperature fluctuations due to the "greenhouse effect" of global warming, melting of

the polar ice caps, and increased severity of seasonal weather conditions.

8. Increasing frequency of natural disasters such as earthquakes, volcanic eruptions, hurricanes, tornadoes, windstorms, ice storms, monsoons and floods.

9. Dwindling world food supplies from increasing drought conditions, combined with continuing problems of global overpopulation.

10. The disappearance of various species of wildlife at increasing rates of extinction.

11. Technological disasters such as Bhopal, India (poison gas), Three Mile Island and Chernobal (nuclear accidents), airplane crashes, electrical blackouts, highrise fires, and collapses of oversized structures such as dams and bridges.

12. Increasingly virulent worldwide health epidemics relating to venereal diseases, AIDS, and related attacks on the immune and reproductive systems of epidemic proportions.

13. Increased acts of terrorism and worldwide violence.

14. The continuation and spread of local wars.

15. The arms race enhancement of nuclear arsenals.

This doomsday list reflects the effect on a global scale of the *either—or* state of consciousness that has existed in the world. Just as the heightened technology of the age of science brought man the material advances of the 20th century, it's abuses have heightened man's sense of separateness and isolation from nature, his fellow human beings and himself—reflecting this overall consciousness of duality.

But what if we were able collectively to expand our awareness out from the limited us versus them consciousness to include the entire boundaries of these two extremes. This expansion would move us from the limitations of this or that into the realm of the "all," where the word *and* becomes the basis of consciousness rather than the word *or*. A place where things become all inclusive rather than mutually exclusive and man is motivated not by selfish considerations of dualism, but by the larger values of unity.

Paradoxically it seems that it has almost been necessary to bring the world to the brink of extinction in order to mobilize the collective consciousness of it's inhabitants to begin to take the necessary action to save it. Everywhere we look nowadays there is heightened awareness and positive action being suggested and taken by all segments of world society as the severity of the problem is recognized. The problems have been identified, we know to a certainty that our planet is in mortal danger, we are aware of many of the causes, and we are starting to zero in on curative action. The question that is asked of course is, "Are we too late?" and "What difference can I make in healing/saving my planet?"

Let's first explore how we arrived at our predicament. The surface causes on the physical plane are the industrial revolution, the urbanization of the population, the dramatic increase in technology and depersonalization of the populous, overpopulation, pollution, wars, genocide, and of course the nuclear spiral of destruction. Beyond all of these specific factors, underlying them and being the real core that feeds their energy are the concepts of separateness and fear—the cosmic opposites of unity and love.

From the cosmic perspective, we are both the causes and the cures for our planetary predicament. Our true identity, of course, is an energy essence that is an integral part of universal conscious awareness. Our individual input into this great reservoir of energy that runs the Universe and translates into the physical experiences in our world of form is crucial. Judging from the fact that the world now seems to be at the crossroad, we must assume our share of responsibility for the predicament. At the same time we can make a difference starting immediately by imputing into the collective energy awareness a heightened consciousness that is manifested in thoughts, words and deeds relating to love and compassion, caring and kindness, and changed belief systems that discard the scenario of Armageddon in favor of returning the earth to Eden. These modified beliefs will, in accordance with our formula for reality creation of awareness-intention-understanding-action, be translated from imagined desire into physical reality.

PRACTICAL ECOLOGY

It's all well and good to speak from the larger perspective in generalities and Universal truths. They are certainly important. However, they must be mated with practical and effective affirming action on the physical plane. So what can we do starting right here and now? What follows is a representative list which is by no means comprehensive, but illustrates what to do and how to do it, and hopefully provides impetus for the reader to "seize the day" and create a miracle:

1. In daily meditation visualize planet earth (you can use that wonderful photograph taken by the lunar astronauts) as a living, breathing organism that needs healing. With the cadence of your breathing fill it with healing white light energy, surround it with a healing force field and visualize it as a viable organism breathing into itself the same energy and expelling with it's out breaths all toxins. Just as you meditate to help the Earth in it's healing, it on a larger scale is a conscious being that, with your love, assistance, and encouragement can also develop it's muscles for self healing.

2. Take into your consciousness visions of doomsday scenarios that are currently manifesting on earth, and then transform those scenes by expanding the negative energy outward and visualizing the opposite reality.

   For example, when you see a polluted stream the water is grey and thick with sediment. Lifeless fish float belly up. There is a strong odor of decay. Then, as you expand your view and elevate your perspective to a higher conscious awareness, the water gradually clears to a shiny, white crystalline clearness. Fish jump happily between the fresh green vegetation. The wind moves the current to and fro. You reach in your cupped hands and drink the cool, clear, refreshing water. It quenches your thirst and fills you with energy and a sense of well being.

3. Honor the earth by everyday actions in your physical plane of existence. This includes not littering, polluting, dirtying, or otherwise fouling your earth home. When you observe people doing so, you ask them, not in a threatening way but from a space of mutual kindness and compassion, to please refrain. You pick up litter and other debris that crosses your path in your normal daily existence and dispose of it properly. You may even wish to organize groups of people in regu-

lar outings to clean up parks, beaches, causeways, and other public areas that have been blighted.

4. Join environmental, ecological and "Save the Planet" groups that focus on raising individual global consciousness to identify and heal Earth abuses.

5. Become familiar with the Green Movement as both a political ideology and a way to live everyday in higher conscious awareness.

6. Raise the awareness of your personal lifestyle as it relates to eating, purchasing goods in the market place, methods of transportation and related aspects of your every day life that impact your environment and earth consciousness. This could include eliminating the consumption of flesh foods and becoming vegetarian, purchasing degradable items, not purchasing products where earth creatures are killed, such as clothing, accessories, furniture, and any other animal-bird-reptile objects.

7. Work to put an end to sport and trophy hunting and fishing. Other than the legitimate food needs of native peoples for indigenous wildlife, there should be no killing whatsoever of these creatures.

8. Join economic boycotting of companies that are contributing to the environmental, ecological and nuclear ills. This includes gasoline and petroleum, food, clothing, automobiles, airplanes, guns and munitions, industrial chemicals and wastes, building materials, and related areas of impact.

9. Encourage organized media attention to the areas of global illness that need healing. Indicate your concern and support for positive programs of awareness and constructive action by letters, phone calls, and any other communication to national and local politicians, magazines, newspapers, television and radio stations, and local community media. This can also involve creating and joining local community action groups to foster and promote awareness of the problems and positive actions to solve them.

10. Expand designated areas of wilderness protection, both nationally and internationally to offer heightened protection

for endangered species, improve and extend their habitat, and at the same time focus attention on the necessity to limit the incursions of man's runaway growth and property development which acts to their detriment.

11. Become more humane in dealing with all creatures. This includes areas of medical and scientific experimentation; zoos, aquariums, artificial recreational captivity problems; and issues of urban domestication.

12. Make monetary contributions wherever possible to promote organizations that are implementing these positive programs. This is the concept on a global level of "Green Energy" which, when combined with the "Energy of Activism" can bring about sudden and dramatic changes for the better in the areas on which attention is focused.

13. Read "Planethood" by Benjamin B. Ferencz and Ken Keyes. It presents in clear and practical terms eight steps we can all follow to alleviate the threat of nuclear catastrophe and assure global peace and prosperity. (Order from Vision Books, 700 Commercial Ave., Coos Bay, Oregon 97420—$2.50)

14. Read "50 Simple things You Can Do To Save The Earth" by The EarthWorks Group. Just published, it offers everyone the opportunity to understand the nature of our planetary predicament and provides daily action available to all that will make a positive difference. (Order from EarthWorks Press, 1400 Shattuck Ave., Box 25, Berkely, California 94709—$4.95)

EARTH DAY

There are many things currently being done to inform, focus attention, and provide a positive forum for conscious action. An example of a specific organized effort is Earth Day 1990 which took place April 22, 1990 to mark the beginning of a long-term commitment to a new sense of responsibility for people, working individually and collectively, to heal and protect the planet.

In April of 1970 hundreds of thousands of people from all walks of life came together to do their individual parts to help clean up the environment. Resulting changes in attitudes and legislation have been significant. Now, 20 years later, this elevated consciousness was organized into a global event.

The concept of "Earth Day" was to launch a decade of the environment which will promote human health, allow biological diversity, and encourage regenerative agriculture. It's focus was to re-educate us on environmentally sound products, policies and investments, some of which include the following:

A. Protecting endangered wildlife.
B. Minimizing hazardous wastes and reducing source violations.
C. Increasing existing energy efficiency and encouraging rapid transition to non-invasive renewable energy resources.
D. Preserving existing old growth forests, curtailing abusive practices, and encouraging effective reforestation techniques.
E. Implementing effective recycling programs and banning packaging that is neither recyclable nor biodegradable.
F. Slowing the rate of global warming by immediate and sustained reductions in automobile and industrial emissions.
G. Creating a world-wide ban on chlorofluoro-carbons to preserve and hopefully reverse existing damage to the ozone layer.
H. Encouraging awareness, responsibility and action by individuals, communities and nations for planetary preservation.

MARKETING ENVIRONMENTALISM

Saving the environment has not only reached a point of critical mass in the collective awareness of our global predicament, but it also is becoming big business. Ironically, much of the same aspects of our technology that created the problem are now being used to cure it. The catalysts are both the dire consequences of failure and the good old profit motive of our Western society. Cleaning up the world is not only the right thing to do, there's also lot's of money to be made in the process.

It is estimated that the environmental services market in the United States will double by 1992 to $108 billion in revenues. The environmental sector has become a classic growth industry, fueled by fear and the growing politics of governmental ecological priorities. Current surveys show a majority of Americans support cleaning up the environment at any cost, and tougher government legislative oversights are providing an impetus to proceed. In addition, the private sector's own punitive mechanism—the lawsuit —adds to this unlikely joinder of interests that are promoting environmentalism.

The basic industry categories of businesses that act "after the fact" to cure abuses are:

1) Handling of existing solid waste
2) Removal and disposal of hazardous wastes
3) Cleanup of pre-existing hazardous wastes
4) Asbestos abatement
5) Recycling of wastes to energy sources
6) Thermal incineration
7) Compliance with EPA regulations
8) Consultation, engineering and equipment supply and maintenance.

We as individuals have also increased our awareness of these ecological problems, and are learning how each of us can make a difference by pursuing environmentally sound actions in our daily lives. This is a massive process of re-education to a "before the fact" awareness, creating a heightened sensitivity to environmental impact, after years of ignorance and neglect. We are unlearning the old ways, and in the process re-creating a more positive reality for ourselves and our world.

One of the best examples of a constructive interrelationship of information and action between the private and business sector is a company called "Seventh Generation" which sells products for a healthy planet through mail order. Their catalogue provides a wealth of information that enables one to both understand environmental issues and modify existing patterns of consuming into a more conscious way of living. They support their commitment to the environment by donating 1% of their gross product sales to non-profit organizations working for a cleaner, healthier planet. (Seventh Generation, 10 Farrell Street, South Burlington, Vermont 05403 1-800-456-1177)

Their main areas of emphasis are:

1) Recyclable paper and wastes

2) Biodegradable plastic and diapers
3) Environmentally safe household cleaners
4) Pesticide-free organic food
5) Water and energy conservation
6) Waste disposal
7) Education

THE GREEN MOVEMENT

The heightened global awareness of environmental issues is a reflection of our overall elevation of consensus consciousness. The idea of re-connecting with all of Nature and constructively reforming our relationship with ourselves and our world has become both a political ideology and a spirit of living day to day, as embodied in "The Greens." Reprinted with permission from the Cascadia Green Alliance, Box 71001, Seattle, Washington 98107, are:

101 Things You Can Do To Promote Green Values

1. Recycle paper, glass and metals
2. Recycle motor oil, dispose of hazardous waste responsibly
3. Use cloth diapers
4. Reuse egg cartons and paper bags
5. Avoid using styrofoam
6. Avoid disposable plates, cups & utensils
7. Use rags instead of paper towels
8. Use paper bags, not paper towels to drain grease
9. Give away rather than dispose of unneeded items
10. Use the back of discardable paper for scratch paper
11. Be responsible and creative with leftover food
12. Use the water from cooking vegetables to make soup
13. Mend and repair rather than discard and replace
14. Invest in well-made, functional clothing
15. Buy bulk & unpackaged rather than packaged goods
16. Purchase goods in reusable or recyclable containers
17. Buy organic, pesticide-free foods
18. Avoid highly processed foods
19. Eat foods from low on the food chain
20. Compost your food scraps

21. Grow your own food (even small kitchen gardens!)
22. Volunteer to start or help with a community garden
23. Support local food co-ops
24. Discover where the food and goods you buy come from
25. Buy locally grown produce and other foods
26. Use glass & steel cookware rather than aluminum
27. Volunteer to maintain local parks & wilderness
28. Buy living Christmas trees
29. Plant trees in your community
30. Learn about the plants & animals in your region
31. Discover your watershed & work to protect it
32. Oppose the use of roadside defoliants in your area
33. Use non-toxic, biodegradable soaps & cleansers
34. Use non-toxic pest control
35. Don't buy products tested on animals
36. Keep hazardous chemicals in spillproof containers
37. Put in a water-conserving shower head
38. Take shorter showers
39. Turn off the water while you brush your teeth
40. Put a water-conservation device in your toilet
41. Learn where your waste & sewage goes
42. Learn where the energy for your home comes from
43. Support your local utility's conservation program
44. Hang your clothes out to dry
45. Be sure your home is appropriately insulated
46. Weather-seal your home thoroughly
47. Heat your home responsibly, with renewable energy
48. Don't burn green wood
49. Choose the long-term investment of solar energy
50. Turn off lights when not in use
51. Turn down your hot water heater
52. Lower your thermostat & wear warmer clothes
53. Buy energy efficient electrical appliances
54. Keep your car engine well tuned
55. Drive a fuel-efficient car that uses unleaded gas
56. Walk, bicycle, carpool or use public transportation
57. Shop by phone first, then pick up your purchases
58. Use rechargeable batteries
59. Research socially-responsible investments
60. Support local credit unions
61. Support local shops & restaurants, not chains
62. "Adopt a grandparent" from the local senior center

63. Volunteer to cook for senior citizens
64. Provide for children in need
65. Hold a community potluck to meet your neighbors
66. Pick up litter along highways & near your home
67. Sponsor a clothes swap
68. Become involved with community projects & events
69. Organize or participate in community sports
70. Be responsible for the values you express
71. Participate in sister city & cultural exchanges
72. Educate yourself on global & "Third World" issues
73. Learn about the cultural diversity of your bioregion
74. Work for global peace
75. Learn how your legislators vote & let them know your views
76. Be an active voter and attend "Town Meetings"
77. Vote for candidates who support Green values
78. Become involved with your child's school
79. Encourage your child's natural talents and interests
80. Organize or join a neighborhood toy co-op
81. Put toxic substances out of reach of children
82. Teach your children ecological wisdom
83. Listen to your child's natural talents and interests
84. Discourage the use of violent toys in your household
85. Communicate openly with your friends & co-workers
86. Acknowledge someone who provides quality service
87. Work to understand people with different views & values
88. Be conscious of the struggles of oppressed people
89. Unlearn cultural sexism and racism
90. Acknowledge the spirituality in yourself and in others
91. Donate blood if your health permits
92. Explore ways to reduce the stress in your life
93. Practice preventive health care
94. Exercise regularly and eat wisely
95. Bring music and laughter into your life
96. Learn about the medication you put into your body
97. Practice responsible family planning
98. Learn First Aid and emergency procedures
99. Take time to play, relax and go into nature
100. Decrease TV watching & increase creative learning
101. HAVE FUN AND BE JOYFUL !!!

So we have alternate scenarios: (1) Doomsday movie which, if allowed to continue unabated, would result in the end of our physical species and our world as we know it, or (2) the Eden movie where we can awaken from our consensus sleep that has brought us to the brink of the precipice, focus our higher conscious awareness with laser beam intensity, and save ourselves and our planet. Our world drama is reminiscent of archetypal story lines where the heroine has been tied to the railroad tracks by the dastardly villain and is rescued by the handsome hero at the last minute, just as the roaring train approaches. They ride off together into the sunset and live happily ever after. Of course if he arrived 10 seconds later the result would have been considerably different. We must act now, not only on the spiritual level where there is no linear time or space, but on the physical level of practical action where we do not have one moment to lose.

IS MAN AN ENDANGERED SPECIES?

Consideration of our global predicament gave me a science fiction story idea which may be closer to possible reality then any of us think. It is presented as it appeared in the copyrighted story outline that is currently under consideration by various major film studios. Hopefully it will only be actualized as a movie for entertainment and informational purposes rather than becoming our experience of physical reality. We, of course, are it's cause and can be it's solution.

Story Outline
THE 23rd CHROMOSOME
Original Idea and Story by Martin E. Segal
Copyright 1988

CHROMOSOME (krō-mō-sōm). "A rod like structure consisting of genes, found in the nuclei of cells. People typically have 23 pairs of chromosomes, the 23rd of which (x or y) determines the sex of the child."

Synopsis: Man's abuse of the Earth produces the ultimate aberration: The human reproductive process yields only female chromosomes—the birth rate of males ceases—human beings become extinct, the ultimate endangered species.

I.   Gaia (the Earth) is dying—due to man's failure to honor it as a living organism and heed the signs of imminent chaos.
II. The Earth has become increasingly toxic due to:
     A. Overpopulation;
     B. Pollution of air, water and food sources;
     C. Interference with the natural order of nature
     D. Man's inhumanity to man.
III. The Earth is in its death throes, and various aspects of nature's perfect mechanisms start to fail one by one. It begins to shut itself down, further aggravating the contributing factors, increasing their severity and ultimate impact.
IV. A mutation occurs that is induced by the increased poisoning of our environment which permanently alters the genetic structure of the male (xy) chromosome so that all human births are female (xx).

A) The scene is a college auditorium. The distinguished professor of biology is answering a hypothetical question posed by one of the many young students in this required course.

Question (very passionately), "Professor, I am a member of the Sierra Club and I deplore the killing of one species after another of our wildlife due to man's abuses. Is it possible for all creatures to eventually become extinct?"

(Impatient groans from the students)

Audience—"That's all we need, another environmental idealist." "Oh, brother!—Set up the soap-box—it's lecture time."

Professor—"Now calm down everyone. (Smiling condescendingly) Let me try to answer the young man's question. Mr. ?"

Student—"Segal."

Audience—"Are you including seagulls in your question?" More laughter.

Professor—"Calm down!—Mr. Segal, let's carry your question a step further to show that your concern, while certainly compassionate, is hardly realistic. Let's see how, from a biological standpoint, man could become extinct—a dinosaur so to speak."

(More laughter from the audience)

Professor—"Now we all know, hopefully from our studies in introductory biology—which you all have presumably passed to arrive here,—(More laughter from the audience.) The definition of a chromosome: At the time of fertilization, the chromosomes from the sperm unite with the chromosomes from the ovum. This union, which is always random, determines the sex of the embryo. The female sex chromosome from the ovum may contribute only an x to the embryo. The male sex chromosome, however, may contribute an x or a y to join with the female chromosome. Thus this union may result in an x x pairing in the sex chromosome, in which case a female will develop; or x y in which case a male will develop."

Professor continues—"This is all basic biology that we should all be familiar with. Now, what if, due to our increasingly dangerous levels of pollution, radiation, atmospheric deterioration, and related toxicity of our environment there occurs an "induced mutation" that permanently alters the genetic structure of the male chromosome so that it becomes identical to the female and can only contribute an x to the embryo? If this occurs, all pairings would be x x, and all births would be female. Eventually, males would become extinct and, of course, that would eventually lead to the extinction of females and the end of the human race as we know it."

(The audience gasps—hushed—absolute silence.)

Professor—"Of course, this could only happen in the movies . . ."

(The audience erupts in nervous laughter, chattering—"Yeah—They could call it 'Attack of the Killer Chromosomes'." (More laughter)

Professor continues—". . . it could not happen in real life!"

B) In a maternity ward in one of the major hospitals, the nurses are looking at statistics of this month's births. There is a very high incidence of female births. They check prior months and request a computer run for the past 12 months to try to explain this unusual occurrence. Similar scenes occur in maternity wards from other hospitals around the world.

C) The statistical data received from the various hospitals, which shows this unmistakable trend toward fewer male births and unusually high female birth rate finds its way to the headquarters of the American Medical Association and related health organizations where the question is asked: "Is this a random occurrence or have we discovered something more suggestive of serious consequences?"

Press conferences are called by the AMA, obstetrician organizations, and related groups to reveal the unusual statistics.

The Surgeon General's office of the Federal Government is called in to review and analyze the data—the implications are enormous!

D) The national media—newspapers, magazines, radio and television become aware of the situation and present various speculations regarding the information's significance and its ultimate consequences. These presentations include the following:

1. Editorial comment by media personalties
2. The opinions of leading scientists, geneticists, biologists, and physicians who seek to analyze the physiological and pathological aspects.
3. The viewpoints of organized religious groups and leaders, including fundamentalists, who state that this is another evidence of God's wrath in the form of plague upon the male offspring similar to Biblical times.
4. Psychologists present viewpoints attempting to analyze the situation as a behavioral aberration. They explain that the increased anxiety and stress of living in our complicated world has temporarily muted the production of male hormones and blocked the "y" chromosome.
5. Philosophers suggest that this is simply a reflection of the decline and fall of our modern society with appropriate references to ancient Greece and the Roman Empire.
6. Environmentalists and ecologists suggest that the earth is a living organism which has been neglected to a point where it is in a critical condition. If, in fact, this condition progresses to a terminal state then, in fact, it is possible to consider a scenario that includes the extinction of the species.
7. New age thinkers and metaphysicians state that all the world's foremost channels have agreed to cooperate in this time of great need, to get together and form a panel of experts

to receive information from higher consciousness and more advanced realms of intelligence concerning the true significance of the situation and the nature of its ultimate possibilities.

8. Feminists and other women's groups proclaim the situation to be their ultimate victory.

E) The linear consequences of the reduction and ultimate cessation of male births are:

1. There will be no new male offspring—males will eventually die off.

2. New births require increased sexual activity of the remaining males, both by actual and artificial insemination. Experimentation with male animal sperm is tried, but fails.

3. Existing sperm bank inventories of "y" sex chromosomes become priceless and are appropriated by the Federal Government in the interests of national security.

4. The balance of power in the economic, social and political structures of the world shifts from male to female. Females achieve psychological and numerical superiority. When they attempt to assume power in the various governments around the world men resist, numerous local wars occur and many males are arrested for political agitation or are killed in the armed conflicts—thus adding to the problem and hastening their extinction.

5. Traditional sex roles and stereotypes are reversed—females assume more and more power, authority and responsibility with attendant increases in stress and anxiety.

6. Male history is erased by the female dominated society. The new society of females publicizes the many abuses of the historical male domination of patriarchal society and vows to create a better world. Hopes rise that all will work out for the best, and in the enthusiasm and euphoria of this "new female order" the somber and lethal real consequences are temporarily overlooked.

F) Once the last males die and the "new female order" is firmly in power, the realization occurs that ultimately all females will incur the same fate. Since no male sperm will exist for conceiving new embryos, there will be no new births. Society will age itself out of existence. The ultimate irony then occurs—women search the globe for any males. From the highest mountain top to the most inaccessible cave, the primary objective becomes the locating of any remaining specimens of the human male species, so new births can occur.

G) The search for males is temporarily successful. Reclusive sects of buddhist monks are located in the Himalayas, Australian aborigines are found in the Outback, Jivaro head hunters are located in the jungles of Ecuador, and Pygmy tribes are found in the Rain Forests of New Guinea. However, this is only temporary and it is impossible to forestall the inevitable.

H) *Conclusion:* Great silence envelops the globe—just like the beginning of the world. There are no human beings. Curiously enough the other life forms have also perished, from birds and animals to the most primitive forms of marine life. The Creator shakes his head, smiles and says,
"Well, it's time to start all over again. Maybe this time they'll get it right!"

A PERSONAL POINT OF VIEW

The individual and collective energies are quickening. We can feel the acceleration of impending change and, while it may become unsettling and we may experience a period of disorientation, we nevertheless know on an intuitive level that cosmic decisions have been made. My view is that our scenario will be Eden and the turbulence that we are experiencing in our everyday lives is evidence that decisive action is being taken toward this end.

The clues are everywhere. Look at the unprecedented world events that have occurred within the last year to raise the banner for peace, understanding, and the elevation of global consciousness:

    1. The Nobel Peace Prize was awarded in 1989 to the Dalai Lama, Tibet's exiled spiritual and political leader. Like Gandhi, he lives a life which reflects his message of universal

reverence and respect for all living things and the advocation of achieving world peace through non-violent means. He points out, "Developing such attitudes as love and compassion, patience and understanding between human beings is not merely a source of personnel happiness, but has become a condition for human survival."

2. The glasnost and perestroika policies of Mikhail Gorbachev are having world wide ramifications in the arena of global politics in Eastern Europe, Asia and Latin America—where long-time policies of repression, separation, and nuclear proliferation are being dramatically reversed.

3. Numerous nations around the world are emerging from military-combative ideologies into atmospheres of increased communication, free discussion, and consideration of human rights.

4. The Berlin Wall, which stood literally and figuratively for 28 years as a symbol of the negative energies of oppression was torn down and the borders between East and West Germany opened in November of 1989. Re-unification, once a dream, became a reality.

5. The government in Romania, the most totalitarian state in the world for the last 25 years, was overthrown in December of 1989, and a more democratic regime of personal liberties established.

6. The great powers of the world are increasing their dialogues for better understanding, mutual cooperation, lessening of military threats through disarmament, and the mutual betterment of their citizens.

7. Concern for nature, the ecological environment, wildlife, and it's critical balance and interrelationship with man is growing daily with emphasis increasing from mere words and theories into effective social action.

8. The creative, formative and causative links between body, mind and spirit are being recognized more and more by the medical profession, science, and New Age spirituality. A linkage is occurring between all, suggesting the greater possibilities of existence, both on the physical and non-physical planes. As our awareness is heightened and our consciousness elevated, both on an individual, community and global

level, we automatically pierce the illusions of separateness that have divided us and emerge into the dawn of a new year.

When New Age Publishing Company was formed in 1975, it's motto and my vision were that,
> "We are embarked upon a new age of insight, awareness, cooperation and growth. We have great dreams—and the opportunity to make them come true."

That vision is being actualized here and now on a global basis. Our future can reflect our true underlying spiritual essence of unconditional love, joy and happiness so that, as I finish this book during Christmas-New Years, our actualized experience of physical reality can be "peace on earth, good will toward men."

What after all is the Holiday season, when stripped of the religious differences that promote separation rather than unity. It is a special time of the year when we emphasize all the best qualities of existence—kindness, caring, compassion, love, helping, joy, sharing, selflessness and generosity. It is a time when, for a few magic moments we are all interconnected as one great family of beings, and there is truly only one of us. Our minds are quiet, our hearts are open, and we enter the realm of our true energy essence conscious awareness. We balance our humanity and divinity and experience unconditional love and higher consciousness. Wouldn't it be wonderful if we could live our everydays in this spirit? This is the true New Age spirituality—my hope for the future. I am reminded of those prophetic words of Robert Frost—

> "The woods are lovely, dark and deep,
> But I have promises to keep,
> And miles to go before I sleep,
> And miles to go before I sleep."

THE END

ABOUT THE AUTHOR

MARTIN E. SEGAL was born in Philadelphia, Pennsylvania and is a long-time resident of Miami, Florida where he is an attorney by vocation, and Adjunct Univ. of Miami Professor.

He is also a student and teacher of Western Psychology, Eastern Philosophy and Higher Consciousness, by avocation.

He lectures on personal growth and spirituality, conducts New Age groups, does community volunteer work, and is the author of *THE GURU IS YOU: How to Play, and Win, the Game of Life* (1985); editor of *AIDS: FROM FEAR TO HOPE — Channeled Teachings Offering Insight and Inspiration* (1987); and wrote the newly published *PEELING THE SWEET ONION: Unlayering the Veils of Identity and Existence* (1990). He is the President of New Age Publishing Co.

THE VERDICT IS IN...

Praise for
THE GURU IS YOU:
How to Play and Win the Game of Life
and
Martin E. Segal

"This book should be able to be read by everyone while offending no one. Highly recommended reading by ANYONE interested in winning ANYTHING!"
—Transformation Times

"This is an interesting, frank, and profound account of how to go beyond the roadblocks that keep us all from enjoying a happy life."
—Ken Keyes, author of the best-selling Handbook to Higher Consciousness

"a sincere and delightful story..."
—Body, Mind Spirit Magazine

"Very inspiring and easy to read."
—M.A., Paradise, CA

"I have already given this book to a friend, thank you for sharing."
—V.H., Catasauqua, PA

"This book is most useful to me as an outline tool for teaching."
—G.B., Louisville, KY

"Title interested me — I knew it was true."
—N.S., Pittsburgh, PA

"Thank you for helping me to help others as well as myself."
J.S., Shreveport, LA

"Your words made me feel good about life and gave me interesting food for thought about the way I play the game."
J.P., No. Miami Beach, FL

"This book will appeal to left brain dominant people, who prefer a well organized presentation of ideas and how to achieve desired results."
S.P., Bixby, OK

"God bless you. Your book The Guru Is You is fantastic. I am ordering two copies, so I can have one to lend".
—K.F., Phoenix, AZ

"A journey to mental health —very informative."
H.M., Weathersfield, CT

"I liked the book very much. Congratulations."
J.M., Tulsa, OK

"Good Stuff, well-shared"
—W.H., Los Angeles, CA

"Excellent." —J.D., No. Little Rock, AR

"A fine piece of work, worthy of wide circulation"
—B.D., Santa Monica, CA

What the Reviewers say about:

AIDS: FROM FEAR TO HOPE

Channeled Teachings Offering Insight and Inspiration

"I can think of only one modern day book of channeled insight that has moved me the way this book has and that is 'A Course in Miracles'. Both have the power to offer not just new understanding, but to illuminate, to turn on a light inside oneself... This is truly a break-through book and deserves to get out into the AIDS mainstream as quickly as possible. It is a significant contribution to a much-needed, expanded, enlightened context for understanding and dealing with AIDS."

—Earth-Star

"The content of the book makes about the most sense I have heard about the gay community... Its central message is that AIDS is calling your attention to spiritual problems: it has a function... This book contains some pearls—"

—Bay Windows

"Regardless of one's opinion of channeling, the material is extremely valuable... the book could not be more credible, for it rings again and again the bells or rightness and truth... It is uplifting, encouraging, often blunt and iconoclastic, always compassionate.

Anyone who has a frightening or terminal condition —which ultimately addresses us all wherever we resist the process of 'life' into 'death'— will find a rich resource of potential guidance in a compatible blend of facts, possibilities and good humor. Read it!"

—New Florida Magazine

"This anthology of channeled comments from various 'Spirit Guides' on the broad spiritual context of AIDS (has a) trans-personal perspective, which includes the subjects of etiology, sexual choice and lifestyle, healing and therapy, and the role of emotions, (and) could be valuable to those who are willing to make the leap of faith."

—EastWest, The Journal of Natural Health & Healing